A STUDENT'S GUIDE TO FOURIER TRANSFORMS
WITH APPLICATIONS IN PHYSICS AND ENGINEERING

J. F. JAMES

Senior Lecturer in Physics
The University of Manchester

CAMBRIDGE
UNIVERSITY PRESS

Published by the Press Syndicate of the University of Cambridge
The Pitt Building, Trumpington Street, Cambridge CB2 1RP
40 West 20th Street, New York, NY 10011-4211 USA
10 Stamford Road, Oakleigh, Melbourne 3166, Australia

First published 1995
Reprinted 1995, 1996

Printed in Great Britain by Athenæum Press Ltd, Gateshead, Tyne & Wear

A catalogue record of this book is available from the British Library

Library of Congress cataloguing in publication data

James, J.F. (John Francis)
A student's guide to Fourier transforms : with applications in
physics and engineering / J.F. James.
p.   cm.
Includes bibliographical references and index.
ISBN 0 521 46298 3 – ISBN 0 521 46829 9 (pbk)
1. Fourier transformations.   2. Mathematical physics.
3. Engineering mathematics.   I. Title.
QC20.7.F67J36   1995
515′.723–dc20   94-22453   CIP

ISBN 0 521 46298 3 hardback

ISBN 0 521 46829 9 paperback

TAG

Contents

Preface

Showing a Fourier transform to a physics student generally produces the same reaction as showing a crucifix to Count Dracula. This may be because the subject tends to be taught by theorists who themselves use Fourier methods to solve otherwise intractable differential equations. The result is often a heavy load of mathematical analysis.

This need not be so. Engineers and practical physicists use Fourier theory in quite another way: to treat experimental data, to extract information from noisy signals, to design electrical filters, to 'clean' TV pictures and for many similar practical tasks. The transforms are done digitally and there is a minimum of mathematics involved.

The chief tools of the trade are the theorems in chapter 2, and an easy familiarity with these is the way to mastery of the subject. In spite of the forest of integration signs throughout the book there is in fact very little integration done and most of that is at high-school level. There are one or two excursions in places to show the breadth of power that the method can give. These are not pursued to any length but are intended to whet the appetite of those who want to follow more theoretical paths.

The book is deliberately incomplete. Many topics are missing and there is no attempt to explain everything: but I have left, here and there, what I hope are tempting clues to stimulate the reader into looking further: and of course, there is a bibliography at the end.

Practical scientists sometimes treat mathematics in general, and Fourier theory in particular, in ways quite different from those for which it was invented†. The late E. T. Bell, mathematician and writer on mathematics, once described mathematics in a famous book title as *The Queen and Servant of Science*. The queen appears here in her role as servant and is

---

† It is a matter of philosophical disputation whether mathematics is invented or discovered. Let us compromise by saying that theorems are discovered: proofs are invented.

sometimes treated quite roughly in that role, and furthermore, without apology. We are fairly safe in the knowledge that mathematical functions which describe phenomena in the real world are 'well-behaved' in the mathematical sense. Nature abhors singularities as much as she does a vacuum.

When an equation has several solutions, some are discarded in a most cavalier fashion as 'unphysical'. This is usually quite right†. Mathematics is after all only a concise shorthand description of the world, and if a position-finding calculation based, say, on trigonometry and stellar observations, gives two results, equally valid, that you are either in Greenland or Barbados, you are entitled to discard one of the solutions if it is snowing outside. So we use Fourier transforms as a guide to what is happening or what to do next, but we remember that for solving practical problems the blackboard-and-chalk diagram, the computer screen and the simple theorems described here are to be preferred to the precise tedious calculations of integrals.

Manchester, January 1994   J. F. James

---

† But Dirac's Equation, with its positive and negative roots, predicted the positron.

1

Physics and Fourier transforms

1.1 The qualitative approach

Ninety per cent of all physics is concerned with vibrations and waves of one sort or another. The same basic thread runs through most branches of physical science, from accoustics through engineering, fluid mechanics, optics, electromagnetic theory and X-rays to quantum mechanics and information theory. It is closely bound to the idea of a *signal* and its *spectrum*. To take a simple example: imagine an experiment in which a musician plays a steady note on a trumpet or a violin, and a microphone produces a voltage proportional to the instantaneous air pressure. An oscilloscope will display a graph of pressure against time, $F(t)$, which is periodic. The reciprocal of the period is the frequency of the note, 256 Hz, say, for a well-tempered middle C.

The waveform is not a pure sinusoid, and it would be boring and colourless if it were. It contains 'harmonics' or 'overtones': multiples of the fundamental frequency, with various amplitudes and in various phases†, depending on the timbre of the note, the type of instrument being played and on the player. The waveform can be *analysed* to find the amplitudes of the overtones, and a list can be made of the amplitudes and phases of the sinusoids which it comprises. Alternatively a graph, $A(v)$, can be plotted (the sound-spectrum) of the amplitudes against frequency.

$A(v)$ **is the Fourier transform of** $F(t)$.

Actually it is the *modular* transform, but at this stage that is a detail.

Suppose that the sound is not periodic – a squawk, a drumbeat or a

---

† 'Phase' here is an angle, used to define the 'retardation' of one wave or vibration with respect to another. One wavelength retardation for example, is equivalent to a phase difference of 2π. Each harmonic will have its own phase, ϕ_m, indicating its position within the period.

1

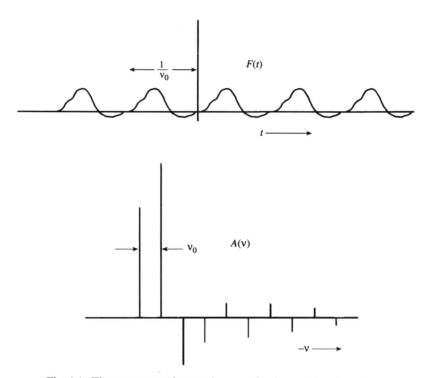

Fig. 1.1. The spectrum of a steady note: fundamental and overtones.

crash instead of a pure note. Then to describe it requires not just a set of overtones with their amplitudes, but a continuous range of frequencies, each present in an infinitesimal amount. The two curves would then look like figure 1.2.

The uses of a Fourier transform can be imagined: the identification of a valuable violin; the analysis of the sound of an aero-engine to detect a faulty gear-wheel; of an electrocardiogram to detect a heart defect; of the light curve of a periodic variable star to determine the underlying physical causes of the variation: all these are current applications of Fourier transforms.

1.2 Fourier series

For a steady note the description requires only the fundamental frequency, its amplitude and the amplitudes of its harmonics. A discrete sum is

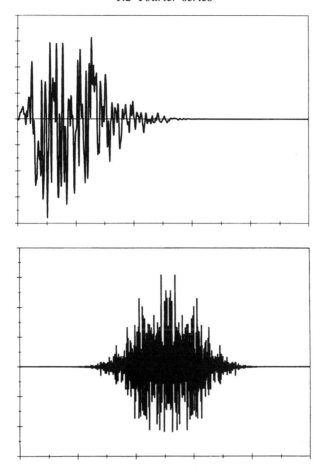

Fig. 1.2. The spectrum of a crash: all frequencies are present.

sufficient. We could write:

$$F(t) \quad = \quad a_0 + a_1 \cos 2\pi v_0 t + b_1 \sin 2\pi v_0 t + a_2 \cos 4\pi v_0 t$$
$$+ b_2 \sin 4\pi v_0 t + a_3 \cos 6\pi v_0 t + \ldots$$

where v_0 is the fundamental frequency of the note. Sines as well as cosines are required because the harmonics are not necessarily 'in step' (i.e., 'in phase') with the fundamental or with each other.

More formally:

$$F(t) = \sum_{n=-\infty}^{\infty} a_n \cos(2\pi n v_0 t) + b_n \sin(2\pi n v_0 t) \qquad (1.1)$$

and the sum is taken from $-\infty$ to ∞ for the sake of mathematical symmetry.

This process of constructing a waveform by adding together a fundamental frequency and overtones or harmonics of various amplitudes, is called *Fourier synthesis*.

There are alternative ways of writing this expression: since $\cos x = \cos(-x)$ and $\sin x = -\sin(-x)$ we can write:

$$F(t) = A_0/2 + \sum_{n=1}^{\infty} A_n \cos(2\pi n v_0 t) + B_n \sin(2\pi n v_0 t) \qquad (1.2)$$

and the two expressions are identical provided that we set $A_n = a_{-n} + a_n$ and $B_n = b_n - b_{-n}$. A_0 is divided by two to avoid counting it twice: as it is, A_0 can be found by the same formula that will be used to find all the A_n's.

Mathematicians and some theoretical physicists write the expression as:

$$F(t) = A_0/2 + \sum_{n=1}^{\infty} A_n \cos(n\omega_0 t) + B_n \sin(n\omega_0 t)$$

and there are entirely practical reasons, which are discussed later, for *not* writing it this way.

1.3 The amplitudes of the harmonics

The alternative process – of extracting from the signal the various frequencies and amplitudes that are present – is called *Fourier analysis* and is much more important in its practical physical applications. In physics, we usually find the curve $F(t)$ experimentally and we want to know the values of the amplitudes A_m and B_m for as many values of m as necessary. To find the values of these amplitudes, we use the *orthogonality* property of sines and cosines. This property is that if you take a sine and a cosine, or two sines or two cosines, each a multiple of some fundamental frequency, multiply them together and integrate the product over one period of that frequency, the result is always zero except in special cases.

If $P, = 1/v_0$, is one period, then:

$$\int_{t=0}^{P} \cos(2\pi n v_0 t) \cdot \cos(2\pi m v_0 t) dt = 0$$

and

$$\int_{t=0}^{P} \sin(2\pi n v_0 t) \cdot \sin(2\pi m v_0 t) dt = 0$$

unless $m = \pm n$, and:

$$\int_{t=0}^{P} \sin(2\pi n v_0 t) \cdot \cos(2\pi m v_0 t) dt = 0$$

always. The first two integrals are both equal to $1/2v_0$ if $m = n$.

We multiply the expression (1.2) for $F(t)$ by $\sin(2\pi m v_0 t)$ and the product is integrated over one period, P:

$$\int_{t=0}^{P} F(t) \sin(2\pi m v_0 t) dt$$

$$= \int_{t=0}^{P} \sum_{n=1}^{\infty} \{A_n \cos(2\pi n v_0 t) + B_n \sin(2\pi n v_0 t)\} \sin(2\pi m v_0 t) dt$$

$$+ \frac{A_0}{2} \int_{t=0}^{P} \sin(2\pi m v_0 t) dt \qquad (1.3)$$

and all the terms of the sum vanish on integration except

$$\int_{0}^{P} B_m \sin^2(2\pi m v_0 t) dt = B_m \int_{0}^{P} \sin^2(2\pi m v_0 t) dt$$

$$= B_m/2v_0 = B_m P/2$$

so that

$$B_m = (2/P) \int_{0}^{P} F(t) \sin(2\pi m v_0 t) dt \qquad (1.4)$$

and provided that $F(t)$ is known in the interval $0 \to P$ the coefficient B_m can be found. If an analytic expression for $F(t)$ is known, the integral can often be done. On the other hand, if $F(t)$ has been found experimentally, a computer is needed to do the integrations.

The corresponding formula for A_m is:

$$A_m = (2/P) \int_{0}^{P} F(t) \cos(2\pi m v_0 t) dt \qquad (1.5)$$

The integral can start anywhere, not necessarily at $t = 0$, so long as it extends over one period.

Example: Suppose that $F(t)$ is a square-wave of period $1/v_0$, so that $F(t) = h$ for $t = -b/2 \to b/2$ and 0 during the rest of the period, as in

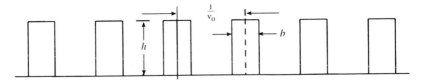

Fig. 1.3. A rectangular wave of period $1/v_0$ and pulse-width b.

the diagram: then

$$A_m = 2v_0 \int_{-1/2v_0}^{1/2v_0} F(t)\cos(2\pi m v_0 t)dt$$

$$= 2hv_0 \int_{-b/2}^{b/2} \cos(2\pi m v_0 t)dt$$

and the new limits cover only that part of the cycle where $F(t)$ is different from zero.

If we integrate and put in the limits:

$$
\begin{aligned}
A_m &= \frac{2hv_0}{2\pi m v_0}\{\sin(\pi m v_0 b) - \sin(-\pi m v_0 b)\} \\
&= \frac{2h}{\pi m}\sin(\pi m v_0 b) \\
&= 2hv_0 b\{\sin(\pi v_0 m b)/\pi v_0 m b\}
\end{aligned}
$$

All the B_n's are zero because of the symmetry of the function – we took the origin to be at the centre of one of the pulses.

The original function of time can be written:

$$F(t) = hv_0 b + 2hv_0 b \sum_{m=1}^{\infty} \{\sin(\pi v_0 m b)/\pi v_0 m b\}\cos(2\pi m v_0 t) \qquad (1.6)$$

or alternatively:

$$F(t) = \frac{hb}{P} + \frac{2hb}{P}\sum_{m=1}^{\infty}\{\sin(\pi v_0 m b)/\pi v_0 m b\}\cos(2\pi m v_0 t) \qquad (1.7)$$

Notice that the first term, $A_0/2$ is the *average* height of the function – the area under the top-hat divided by the period: and that the function $\sin(x)/x$, called 'sinc(x)', which will be described in detail later, has the value unity at $x = 0$, as can be shown using De l'Hôpital's rule†.

† De l'Hôpital's rule is that if $f(x) \to 0$ as $x \to 0$ and $\phi(x) \to 0$ as $x \to 0$, the ratio $f(x)/\phi(x)$ is indeterminate, but is equal to the ratio $(df/dx)/(d\phi/dx)$ as $x \to 0$.

There are other ways of writing the Fourier series. It is convenient occasionally, though less often, to write $A_m = R_m \cos \phi_m$ and $B_m = R_m \sin \phi_m$, so that equation (1.2) becomes:

$$F(t) = \frac{A_0}{2} + \sum_{m=1}^{\infty} R_m \cos(2\pi m v_0 t + \phi_m) \qquad (1.8)$$

and R_m and ϕ_m are the amplitude and phase of the mth harmonic. A single sinusoid then replaces each sine and cosine, and the two quantities needed to define each harmonic are these amplitudes and phases in place of the previous A_m and B_m coefficients. In practice it is usually the amplitude, R_m which is important, since the energy in an oscillator is proportional to the square of the amplitude of oscillation, and $|R_m|^2$ gives a measure of the power contained in each harmonic of a wave. 'Phase' is a simple and important idea. Two wave trains are 'in phase' if wave crests arrive at a certain point together. They are 'out of phase' if a trough from one arrives at the same time as the crest of the other. (Alternatively they have 180° phase difference.) In figure 1.4, there are two wave trains. The upper has 0.7× the amplitude of the other and it *lags* (not *leads*, as it appears to do) the lower by 70°. This is because the horizontal axis of the graph is time, and the vertical axis measures the amplitude at a fixed point as it varies with time. Wave crests from the lower wave train arrive earlier than those from the upper. The important thing is that the 'phase-difference' between the two is 70°.

It is usually more convenient to use complex exponentials to manipulate formulae than to use sines and cosines, simply because they are easier to handle. This tends to make the theory look complicated, since the functions $F(t)$ and $\Phi(v)$ can now be complex functions (of a real variable). This means, for example, that when a real number v is put into $\Phi(v)$, a complex number is obtained. As we shall see, this only refers to the amplitudes of the sine and cosine components, and there is an easy physical interpretation of the complex number.

Thus with complex representation we have:

$$F(t) = \frac{A_0}{2} + \sum_{m=1}^{\infty} e^{2\pi i m v_0 t}(A_m - iB_m)$$

or

$$F(t) = \frac{A_0}{2} + \sum_{m=1}^{\infty} e^{2\pi i m v_0 t} C_m \qquad (1.9)$$

where C_m is now a complex amplitude. The phase angle ϕ_m men-

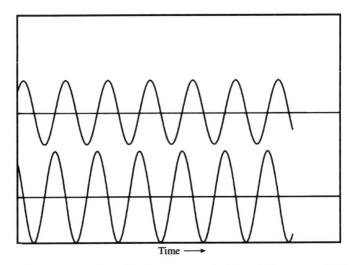

Fig. 1.4. Two wave trains with the same period but different amplitudes and phases. The upper has 0.7× the amplitude of the lower and there is a phase-difference of 70°.

tioned earlier is given by $\tan\phi_m = -B_m/A_m$, and the amplitude, $R_m = \sqrt{A_m^2 + B_m^2}$. These are of course the argument and modulus of the complex number, C_m.

The formal derivations for these versions of the Fourier series are in the appendix. What is important here is that in each case the coefficients A_m and B_m are obtained from the *Inversion Formulae*:

$$A_m = 2v_0 \int_0^{1/v_0} F(t)\cos(2\pi m v_0 t)dt$$

$$B_m = 2v_0 \int_0^{1/v_0} F(t)\sin(2\pi m v_0 t)dt$$

$$C_m = 2v_0 \int_0^{1/v_0} F(t)e^{-2\pi i m v_0 t}dt$$

(note: the minus sign in the exponent is important) or, if ω_0 has been used instead of v_0 $(= \omega_0/2\pi)$ then:

$$A_m = \omega_0/\pi \int_0^{2\pi/\omega_0} F(t)\cos(m\omega_0 t)dt$$

$$B_m = \omega_0/\pi \int_0^{2\pi/\omega_0} F(t)\sin(m\omega_0 t)dt$$

$$C_m = 2\omega_0/\pi \int_0^{2\pi/\omega_0} F(t)e^{-im\omega_0 t}dt$$

The useful mnemonic form to remember for finding the coefficients in a Fourier series is:

$$A_m = \frac{2}{period} \int_{one\ period} F(t)\cos\left\{\frac{2m\pi t}{period}\right\}dt \tag{1.10}$$

$$B_m = \frac{2}{period} \int_{one\ period} F(t)\sin\left\{\frac{2m\pi t}{period}\right\}dt \tag{1.11}$$

and remember that the integral can be taken from any starting point, a, provided it extends over one period to an upper limit $a + P$. The integral can be split into as many subdivisions as needed, if, for example $F(t)$ has different analytic forms in different parts of the period.

1.4 Fourier transforms

Whether $F(t)$ is periodic or not, a complete description of $F(t)$ can be given using sines and cosines. If $F(t)$ is not periodic it requires all frequencies to be present if it is to be synthesised. A non-periodic function may be thought of as a limiting case of a periodic one, where the period tends to infinity, and consequently the fundamental frequency tends to zero. The harmonics are more and more closely spaced and in the limit there is a continuum of harmonics, each one of infinitesimal amplitude, $a(v)dv$, for example. The summation sign is replaced by an integral sign and we find that

$$F(t) = \int_{-\infty}^{\infty} a(v)dv \cos(2\pi vt) + \int_{-\infty}^{\infty} b(v)dv \sin(2\pi vt) \tag{1.12}$$

or, equivalently:

$$F(t) = \int_{-\infty}^{\infty} a(v)\cos(2\pi vt + \phi(v))dv \tag{1.13}$$

or, again:

$$F(t) = \int_{-\infty}^{\infty} \Phi(v)e^{2\pi i v t}dv \qquad (1.14)$$

If $F(t)$ is real, that is to say, if the insertion of any value of t into $F(t)$ yields a real number, then $A(v)$ and $B(v)$ are real too. However, $\Phi(v)$ may be complex and indeed will be if $F(t)$ is asymmetrical so that $F(t) \neq F(-t)$. This can sometimes cause complications, and these are dealt with in chapter 7: but $F(t)$ is often symmetrical and then $\Phi(v)$ is real and $F(t)$ comprises only cosines. We *could* then write :

$$F(t) = \int_{-\infty}^{\infty} a(v)\cos(2\pi v t)dv$$

but because complex exponentials are easier to manipulate, we take as a standard form the equation (1.14) above. Nevertheless, for many practical purposes only real and symmetrical functions $F(t)$ and $\Phi(v)$ need be considered.

Just as with Fourier series, the function $\Phi(v)$ can be recovered from $F(t)$ by inversion. This is the cornerstone of Fourier theory because, astonishingly, the inversion has exactly the same form as the synthesis, and we can write, if $\Phi(v)$ is real and $F(t)$ is symmetric:

$$\Phi(v) = \int_{-\infty}^{\infty} F(t)\cos(2\pi v t)dt \qquad (1.15)$$

so that not only is $\Phi(v)$ the Fourier transform of $F(t)$, but $F(t)$ is the Fourier transform of $\Phi(v)$. The two together are called a 'Fourier pair'.

The complete and rigorous proof of this is long and tedious† and it is not necessary here; but the formal definition can be given and this is a suitable place to abandon, for the moment, the physical variables time and frequency and to change to the pair of abstract variables, x and p, which are usually used. The formal statement of a Fourier transform is then:

$$\Phi(p) = \int_{-\infty}^{\infty} F(x)e^{2\pi i p x}dx \qquad (1.16)$$

$$F(x) = \int_{-\infty}^{\infty} \Phi(p)^{-2\pi i p x}dp \qquad (1.17)$$

---

† It is to be found, for example, in E.C. Titchmarsh, *Introduction to the Theory of Fourier Integrals*, Clarendon Press, Oxford, 1962, or in R.R. Goldberg, *Fourier Transforms*, Cambridge University Press, 1965.

and this pair of formulae† will be used from here on.
Symbolically we write:

$$\Phi(p) \rightleftharpoons F(x).$$

One and only one of the integrals must have a minus sign in the exponent. Which of the two you choose does not matter, so long as you keep to the rule. If the rule is broken half way through a long calculation the result is chaos; but if someone else has used the opposite choice, the Fourier pair calculated of a given function will be the complex conjugate of that given by your choice.

When time and frequency are the conjugate variables we shall use:

$$\Phi(v) = \int_{-\infty}^{\infty} F(t)e^{-2\pi i v t} dt \tag{1.19}$$

$$F(t) = \int_{-\infty}^{\infty} \Phi(v)^{2\pi i v t} dv \tag{1.20}$$

and again, symbolically:

$$\Phi(v) \rightleftharpoons F(t)$$

There are two good reasons for incorporating the 2π into the exponent. Firstly the defining equations are easily remembered without worrying where the 2π's go, but more importantly, quantities like t and v are actually physically measured quantities – time and frequency – rather than time and *angular* frequency, ω. Angular measure is for mathematicians. For example, when one has to integrate a function wrapped around a cylinder it is convenient to use the angle as the independent variable. Physicists will generally find it more convenient to use t and v, for example, with the 2π in the exponent.

† Sometimes one finds:

$$\Phi(p) = \frac{1}{2\pi} \int_{-\infty}^{\infty} F(x)e^{ipx} dx \quad ; \qquad F(x) = \int_{-\infty}^{\infty} \Phi(p)e^{-ipx} dp$$

as the defining equations, and again symmetry is preserved by some people by defining the transform by:

$$\Phi(p) = \left\{ \frac{1}{2\pi} \right\}^{\frac{1}{2}} \int_{-\infty}^{\infty} F(x)e^{ipx} dx \quad ; \qquad F(x) = \left\{ \frac{1}{2\pi} \right\}^{\frac{1}{2}} \int_{-\infty}^{\infty} \Phi(p)e^{-ipx} dp. \tag{1.18}$$

1.5 Conjugate variables

Traditionally x and p are used when abstract transforms are considered and they are called 'conjugate variables'. Different fields of physics and engineering use different pairs, such as frequency, v and time, t in accoustics, telecommunications and radio; position, x and momentum divided by Planck's constant, p/h in quantum mechanics, and aperture x, and diffraction angle divided by wavelength $p = \sin\theta/\lambda$ in diffraction theory.

In general we will use x and p as abstract entities and give them a a physical meaning when an illustration seems called-for. It is worth remembering that x and p have inverse dimensionality, as in time t and frequency t^{-1}. The product px, like any exponent, is always a dimensionless number.

One further definition is needed: the 'Power Spectrum' of a function.† This notion is important in electrical engineering as well as in physics. If power is transmitted by electromagnetic radiation (radio waves or light) or by wires or waveguides, the voltage at a point varies with time as $V(t)$. $\Phi(v)$, the Fourier transform of $V(t)$, may very well be – indeed usually is – complex. However, the power per unit frequency interval being transmitted is proportional to $\Phi(v)\Phi^*(v)$, where the constant of proportionality depends on the load impedance. The function $S(v) = \Phi(v)\Phi^*(v) = |\Phi(v)|^2$ is called the Power Spectrum or the Spectral Power Density (SPD) of $F(t)$. This is what an optical spectrometer measures, for example.

1.6 Graphical representations

It frequently happens that greater insight into the physical processes which are described by a Fourier transform can be achieved by a diagram rather than a formula. When a real function $F(x)$ is transformed it generally produces a complex function $\Phi(p)$, which needs an Argand diagram to demonstrate it. Three dimensions are required: $\mathbf{Re}\Phi(p)$, $\mathbf{Im}\Phi(p)$ and p. A perspective drawing will display the function, which appears as a more or less sinuous line. If $F(x)$ is symmetrical, the line lies in the $\mathbf{Re}$-p plane, and if antisymmetrical, in the $\mathbf{Im}$-p plane.

Electrical engineering students in particular, will recognise the end-on

---

† Actually the *energy* spectrum. 'power spectrum' is just the conventional term used in most books. This is discussed in more detail in chapter 4.

view along the p-axis as the 'Nyquist Diagram' of feedback theory. There will be examples of this graphical representation in later chapters.

1.7 Useful functions

There are some functions which occur again and again in physics, and whose properties should be learned. They are extremely useful in the manipulation and general taming of other functions which would otherwise be almost unmanageable. Chief among these are:

1.7.1 The 'top-hat' function†, $\Pi_a(x)$

This has the property that:

$$
\begin{aligned}
\Pi_a(x) &= 0, -\infty < x < -a/2 \\
&= 1, -a/2 < x < a/2 \\
&= 0, a/2 < x < \infty
\end{aligned}
$$

and the symbol Π is chosen as an obvious aid to memory.

Its Fourier pair is obtained by integration:

$$
\begin{aligned}
\Phi(p) &= \int_{-\infty}^{\infty} \Pi_a(x) e^{2\pi i p x} dx \\
&= \int_{-a/2}^{a/2} e^{2\pi i p x} dx \\
&= \frac{1}{2\pi i p} [e^{\pi i p a} - e^{-\pi i p a}] \\
&= a \left\{ \frac{\sin \pi p a}{\pi p a} \right\} \\
&= a \cdot \text{sinc}(\pi p a)
\end{aligned}
$$

and the 'sinc-function', defined‡ by $\text{sinc}(x) = \sin x / x$ is one which recurs throughout physics. As before, we write symbolically:

$$
\Pi_a(x) \rightleftharpoons a \cdot \text{sinc}(\pi p a)
$$

† In the USA, this is called a 'box-car' or 'rect' function.
‡ Caution: some people define $\text{sinc}(x)$ as $\sin(\pi x)/(\pi x)$.

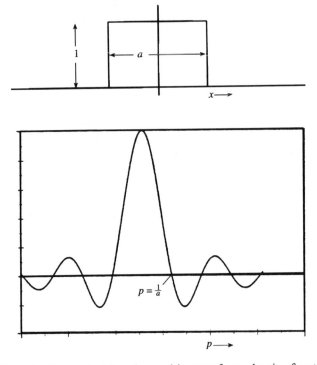

Fig. 1.5. The top-hat function and its transform, the sinc-function.

1.7.2 The sinc-function

$$\text{sinc}(x) = \sin x / x$$

Has the value unity at $x = 0$, and has zeros whenever $x = n\pi$. The function $\text{sinc}(x)$ above, the most common form, has zeros when $p = 1/a, 2/a, 3/a \ldots$

1.7.3 The Gaussian function

Suppose $G(x) = e^{-x^2/a^2}$.

a is the 'width parameter' of the function, and the Full Width at Half Maximum (FWHM) is $1.386a$, and (what every scientist should know!): $\int_{-\infty}^{\infty} e^{-x^2/a^2} dx = a\sqrt{\pi}$.

Its Fourier transform is $g(p)$, given by:

$$g(p) = \int_{-\infty}^{\infty} e^{-x^2/a^2} e^{2\pi i p x} dx$$

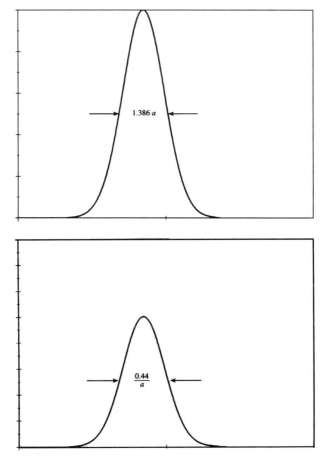

Fig. 1.6. The Gaussian function and its transform, another Gaussian with full width at half maximum inversely proportional to that of its Fourier pair.

The exponent can be rewritten (by 'completing the square') as

$$-(x/a - \pi ipa)^2 - \pi^2 p^2 a^2$$

and then

$$g(p) = e^{-\pi^2 p^2 a^2} \int_{-\infty}^{\infty} e^{-(x/a - \pi ipa)^2} dx$$

put $x/a - \pi ipa = z$, so that $dx = adz$. Then:

$$
\begin{aligned}
g(p) &= ae^{-\pi^2 p^2 a^2} \int_{-\infty}^{\infty} e^{-z^2} dz \\
&= a\sqrt{\pi} e^{-\pi^2 a^2 p^2}
\end{aligned}
$$

so that $g(p)$ is another Gaussian function, with width parameter $1/\pi a$.

Notice that, the wider the original Gaussian, the narrower will be its Fourier pair.

Notice too, that the value at $p = 0$ of the Fourier pair is equal to the area under the original Gaussian.

1.7.4 The exponential decay

This, in physics is generally the positive part of the function $e^{-x/a}$. It is asymmetric, so its Fourier transform is complex:

$$\Phi(p) = \int_0^\infty e^{-x/a} e^{2\pi i p x} dx$$

$$= \left[\frac{e^{2\pi i p x - x/a}}{2\pi i p - 1/a} \right]_0^\infty = \frac{-1}{2\pi i p - 1/a}$$

Usually, with this function, the power spectrum is the most interesting:

$$| \Phi(p) |^2 = \frac{a^2}{4\pi^2 p^2 a^2 + 1}$$

This is a bell-shaped curve, similar in appearance to a Gaussian curve, and is known as a Lorentz Profile. It is the shape found in spectrum lines, $I(v)$, when they are observed at very low pressure, and the FWHM, Δv, which is $1/\pi a$, is related to the 'Lifetime of the Excited State', the reciprocal of the 'transition probability', of the atom which produces the transition. In this example, a and x obviously have dimensions of time.

1.7.5 The Dirac 'delta-function'

This has the following properties:

$\delta(x) = 0$ unless $x = 0$

$\delta(0) = \infty$

$\int_{-\infty}^\infty \delta(x) dx = 1$

It is an example of a function which disobeys one of Dirichlet's conditions, since it is unbounded at $x = 0$. It can be regarded crudely as the limiting case of a top-hat function $(1/a)\Pi_a(x)$ as $a \to 0$. It becomes narrower and higher, and its area is always equal to unity. Its Fourier transform is sinc$(\pi p a)$ and as $a \to 0$, sinc$(\pi p a)$ stretches and in the limit

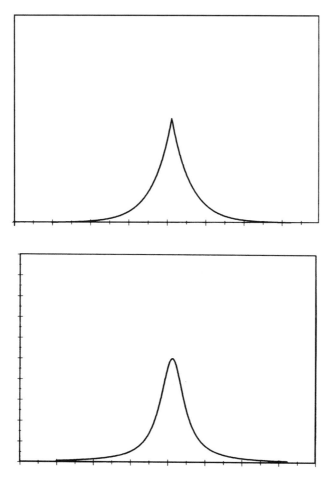

Fig. 1.7. The exponential decay $e^{-|x|/a}$ and its Fourier Transform.

is a straight line at unit height above the x-axis. In other words:

The Fourier transform of a δ-function is unity

and we write:

$$\delta(x) \rightleftharpoons 1$$

Alternatively, and more accurately, it is the limiting case of a Gaussian function of unit area as it gets narrower and higher. Its Fourier transform then is another Gaussian of unit height, getting broader and broader until in the limit it is a straight line at unit height above the axis.

Its usefulness will become apparent when we consider the shift theorems.

Other useful properties to memorise are:

$$\delta(x - a) = 0 \text{ unless } x = a$$

$$\int_{-\infty}^{\infty} F(x)\delta(x - a)dx = F(a)$$

and as a consequence of this:

$$\int_{-\infty}^{\infty} e^{2\pi ipx}\delta(x - a)dx = e^{2\pi ipa}$$

so that we can write:

$$\delta(x - a) \rightleftharpoons e^{2\pi ipa}$$

1.7.5.1 A pair of δ-functions

If two δ-functions are equally disposed on either side of the origin, the Fourier transform is a cosine wave:

$$\delta(x - a) + \delta(x + a) \quad \rightleftharpoons \quad e^{2\pi ipa} + e^{-2\pi ipa} \qquad (1.21)$$
$$= \quad 2\cos(2\pi pa)$$

1.7.5.2 The Dirac Comb

This is an infinite set of equally-spaced δ-functions, usually denoted by the Cyrillic letter III (shah). formally, we write:

$$III_a(x) = \sum_{n=-\infty}^{\infty} \delta(x - na)$$

It is useful because it allows us to include Fourier series in the general theory of Fourier transforms. For example, the *convolution* (to be described later) of $III_a(x)$ and $(1/b)\Pi_b(x)$ (where $b < a$) is a square wave similar to that in the earlier example, of period a and width b, and with unit area in each rectangle. The Fourier transform is then a Dirac comb, with 'teeth' of height a_m spaced at intervals $1/a$. The a_m are of course the coefficients in the series.

If the square wave is allowed to become infinitesimally wide and infinitely high so that the area under each rectangle remains unity, then the coefficients a_m will all become the same height, $1/a$. In other words, the Fourier transform of a Dirac comb is another Dirac comb:

$$III_a(x) \rightleftharpoons \frac{1}{a}III_{\frac{1}{a}}(p)$$

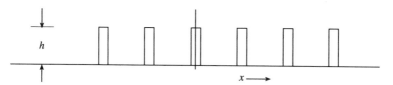

Fig. 1.8. A rectangular pulse-train with a 4:1 'mark–space' ratio.

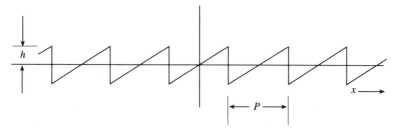

Fig. 1.9.

and again notice that the period in p-space is the reciprocal of the period in x-space.

This is not a formal demonstration of the Fourier transform of a Dirac comb. A rigorous proof is much more elaborate, but is unnecessary here.

1.8 Worked examples

(1) A train of rectangular pulses has a pulse width equal to 1/4 of the pulse period. Show that the 4th, 8th, 12th etc. harmonics are missing.

Taking zero at the centre of one pulse, the function is clearly symmetrical so that there are only cosine amplitudes.

$$
\begin{aligned}
A_n &= \left(\frac{2}{P}\right) \int_{-P/8}^{P/8} h \cos\left(\frac{2\pi nx}{P}\right) dx \\
&= \left(\frac{h}{\pi n}\right) 2 \sin\left(\frac{2\pi n}{P} \cdot \frac{P}{8}\right) \\
&= \left(\frac{h}{2}\right) \operatorname{sinc}\left(\frac{\pi n}{4}\right)
\end{aligned}
$$

so that $A_n = 0$ if $n = 4, 8, 12, \ldots$

(2) Find the sine-amplitude of a saw-tooth waveform as in figure 1.9.

By choosing the origin half way up one of the teeth, the function is clearly made antisymmetrical, so that there are no cosine amplitudes.

$$
\begin{aligned}
B_n &= \frac{2}{P} \int_{-P/2}^{P/2} \frac{2xh}{P} \sin\left(\frac{2\pi nx}{P}\right) dx \\
&= \frac{4h}{P^2} \left[-x\cos\left(\frac{2\pi nx}{P}\right) \frac{P}{2\pi n} + \frac{P^2}{4\pi^2 n^2} \sin\left(\frac{2\pi nx}{P}\right) \right]_{-P/2}^{P/2} \\
&= (-2h/\pi n)\cos\pi n \quad \text{since } \sin\pi n = 0
\end{aligned}
$$

so that
$$
B_n = -(-1)^n (2h/\pi n)
$$

As a matter of interest, it is worth while calculating the sine-amplitudes when the origin is taken at the tip of a tooth, to see how changing the position of the waveform changes the amplitudes. It is also worth while doing the calculation for a similar wave, with negative-going slopes instead of positive.

2

Useful properties and theorems

2.1 The Dirichlet conditions

Not all functions can be Fourier-transformed. They are transformable if they fulfil certain conditions, known as the Dirichlet conditions.

The integrals which formally define the Fourier transform in chapter 1 will exist if the integrands fulfil the following conditions:

The functions $F(x)$ and $\Phi(p)$ are square-integrable, e.g. $\int_{-\infty}^{\infty} |F(x)|^2 \, dx$ is finite which implies that $F(x) \to 0$ as $|x| \to \infty$

$F(x)$ and $\Phi(p)$ are single-valued. For example a function such as that in figure 2.1 is not Fourier-transformable:

$F(x)$ and $\Phi(p)$ are 'piece-wise continuous'. The function can be broken

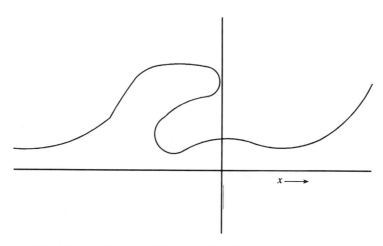

Fig. 2.1. A triple-valued function like this can not be Fourier-transformed.

21

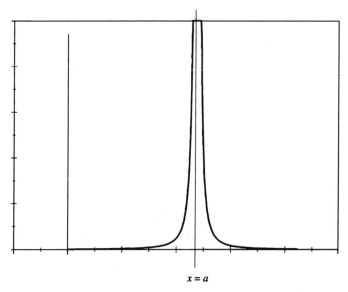

$x = a$

Fig. 2.2. $F(x) = 1/(x-a)^2$, an unbounded non-transformable function of x.

up into separate pieces, so that there can be isolated discontinuities, as many as you like, at the junctions, but the functions must be *continuous* in the mathematical sense, between these discontinuities.†

The functions $F(x)$ and $\Phi(p)$ have upper and lower bounds. This is a condition which is *sufficient* but has not been proved *necessary*. In fact we shall assume that it is not. The Dirac δ-function, for instance, disobeys this condition. No engineer or physicist has yet lost sleep over this one.

In Nature, all the phenomena that can be described mathematically seem to require only well-behaved functions which obey the Dirichlet conditions. For example, we can desribe the electric field of a wave-packet‡ by a function which is continuous, finite and single-valued everywhere,

---

† The classical nonconformist example is Weierstrass' function, $W(x)$, which has the property that $W(x) = 1$ if x is rational and $W(x) = 0$ if x is irrational. It looks like a straight line but it is not transformable, since it can be shown that between any two rational numbers, however close, there is at least one irrational number, and between any two irrational numbers there is at least one rational number, so that the function is everywhere discontinuous.

‡ I have deliberately avoided the word 'photon', for fear of causing apoplexy among strict quantum theory purists.

and as the wave-packet contains only a finite amount of energy, the electric field is square-integrable.

2.2 Theorems

There are several theorems which are of great use in manipulating Fourier-pairs, and they should be memorised. For the most part the proofs are elementary. The art of practical Fourier-transforming is in the manipulation of functions using these theorems, rather than in doing extensive and tedious elementary integrations. It is this, as much as anything, which makes Fourier theory such a powerful tool for the practical working scientist.

In what follows, we assume:

$$F_1(x) \rightleftharpoons \Phi_1(p) \; ; \; F_2(x) \rightleftharpoons \Phi_2(p)$$

where '$\rightleftharpoons$' implies that F_1 and Φ_1 are a Fourier pair.

The Addition Theorem:

$$F_1(x) + F_2(x) \rightleftharpoons \Phi_1(p) + \Phi_2(p) \tag{2.1}$$

The Shift Theorems:

$$F_1(x + a) \rightleftharpoons \Phi_1(p)e^{2\pi ipa}$$

$$F_1(x - a) \rightleftharpoons \Phi_1(p)e^{-2\pi ipa} \tag{2.2}$$

$$F_1(x - a) + F_1(x + a) \rightleftharpoons 2\Phi_1(p)\cos 2\pi pa$$

The third of these theorems can be illustrated: In particular, notice that if $F_1(x)$ is the δ-function, the shift theorems are:

$$\delta(x + a) \rightleftharpoons e^{-2\pi ipa}$$

$$\delta(x - a) \rightleftharpoons e^{2\pi ipa} \tag{2.3}$$

$$\delta(x - a) + \delta(x + a) \rightleftharpoons 2\cos 2\pi pa$$

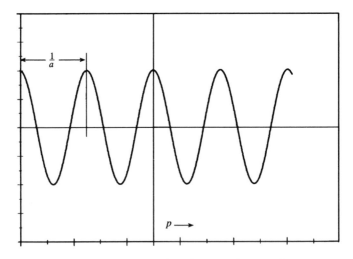

Fig. 2.3. A pair of δ-functions and its transform.

2.3 Convolutions and the Convolution Theorem

Convolutions are an important concept, especially in practical physics, and the idea of a convolution can be illustrated simply by an example.

Imagine a 'perfect' spectrometer, plotting a graph of intensity against wavelength, of a monochromatic source of light of intensity S and wavelength (λ_0). Represent the Power Spectral Density ('the spectrum') of the source by $S\delta(\lambda_0 - \lambda)$. The spectrometer will plot the graph as $kS\delta(\lambda_0 - \lambda)$, where k is a factor which depends on the throughput of the spectrometer, its geometry and its detector sensitivity. No spectrometer is perfect in practice, and what a real instrument will plot in response to a monochromatic input is a continuous curve $kSI(\lambda_0 - \lambda)$, where $I(\lambda)$ is called the 'instrumental function' and $\int I(\lambda)d\lambda = 1$.

Now we inquire what the instrument will plot in response to a continuous spectrum input. Suppose that the intensity of the source as a

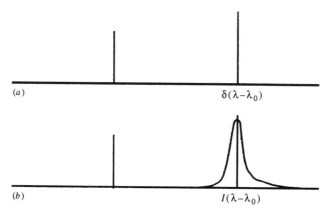

Fig. 2.4. The spectrum of a monochromatic wave (a) entering and (b) leaving a spectrometer. The area under curve (b) must be unity – the same as the 'area' under the δ-function, to preserve the idea of an 'instrumental function'.

function of wavelength is $S(\lambda)$. We assume that a monochromatic line at *any* wavelength λ_1 will be plotted as a similarly shaped function $I(\lambda_1 - \lambda)$. Then an infinitesimal interval of the spectrum can be considered as a monochromatic line, at λ_1, say, and of intensity $S(\lambda_1)d\lambda_1$ and it is plotted by the spectrometer as a function of λ:

$$dO(\lambda) = kS(\lambda_1)d\lambda_1 I(\lambda_1 - \lambda)$$

and the intensity *apparently* at another wavelength λ_2 is

$$dO(\lambda_2) = kS(\lambda_1)I(\lambda_1 - \lambda_2)d\lambda_1$$

The total power apparently at λ_2 is got by integrating this over all wavelengths:

$$O(\lambda_2) = k \int_{-\infty}^{\infty} S(\lambda_1)I(\lambda_1 - \lambda_2)d\lambda_1$$

or, dropping unnecessary subscripts:

$$O(\lambda) = k \int_{-\infty}^{\infty} S(\lambda_1)I(\lambda_1 - \lambda)d\lambda_1$$

and the output curve, $O(\lambda)$ is said to be the convolution of the spectrum $S(\lambda)$ with the instrumental function $I(\lambda)$.

It is the idea of an instrumental function, $I(\lambda)$, which is important here. We assume that the same shape $I(\lambda)$ is given to any monochromatic line input. The idea extends to all sorts of measuring instruments and has various names, such as 'impulse response', 'point-spread function',

'Green's function' and so on, depending on which branch of physics or electrical engineering is being discussed. In an electronic cicuit, for example, it answers the question 'if you put in a sharp pulse, what comes out?' Most instruments have no fixed unique 'instrumental function', but the function often changes slowly enough (with wavelength, in the spectrometer example) that the idea can be used for practical calculations.

The same idea can be envisaged in two dimensions: a point object – a star for instance – is imaged by a camera lens as a small smear of light, the 'point-spread function' of the lens. Even a 'perfect' lens has a diffraction pattern, so that the best that can be done is to convert a point object into an 'Airy disc' – a spot, $1.22f\lambda/d$ in diameter, where f is the focal length and d the diameter of the lens. The lens in general, when taking a photograph, gives an image which is the convolution, in two dimensions, of its point-spread function with the object.

The formal definition of a convolution of two functions is then:

$$C(x) = \int_{-\infty}^{\infty} F_1(x')F_2(x - x')dx' \qquad (2.4)$$

and we write this symbolically as:

$$C(x) = F_1(x) * F_2(x)$$

Convolutions obey various rules of arithmetic, and can be manipulated using them.

The commutative rule:

$$C(x) = F_1(x) * F_2(x) = F_2(x) * F_1(x)$$

or:

$$C(x') = \int_{-\infty}^{\infty} F_2(x)F_1(x' - x)dx$$

as can be shown by a simple substitution.

The distributive rule:

$$F_1(x) * [F_2(x) + F_3(x)] = F_1(x) * F_2(x) + F_1(x) * F_3(x)$$

The associative rule: The idea of a convolution can be extended to three or more functions, and the *order* in which the convolutions are done does not matter.

$$F_1(x) * [F_2(x) * F_3(x)] = [F_1(x) * F_2(x)] * F_3(x)$$

and usually the convolution of three functions is written without the square bracket:

$$C(x) = F_1(x) * F_2(x) * F_3(x) = \int_{-\infty}^{\infty} \int_{-\infty}^{\infty} F_1(x-x')F_2(x'-x'')F_3(x'')dx'dx''$$

In fact a whole algebra of convolutions exists and is very useful in taming some of the more fearsome-looking functions that are found in physics. For example:

$$[F_1(x) + F_2(x)] * [F_3(x) + F_4(x)] = F_1(x) * F_3(x) + F_1(x) * F_4(x)$$

$$+F_2(x) * F_3(x) + F_2(x) * F_4(x)$$

There is a way of visualising a convolution. Draw the graph of $F_1(x)$. Draw, on a piece of transparent paper, the graph of $F_2(x)$. Turn the transparent graph over about a vertical axis and lay this mirror-image of F_2 on top of the graph of F_1. When the two y-axes are displaced by a distance x', integrate the product of the two functions. The result is one point on the graph of $C(x')$.

2.3.1 The Convolution Theorem

With the exception of Fourier's Inversion Theorem, the Convolution Theorem is the most astonishing result in Fourier theory. It is as follows:

If $C(x)$ is the convolution of $F_1(x)$ with $F_2(x)$ then its Fourier pair, $\Gamma(p)$ is the *product* of $\Phi_1(p)$ and $\Phi_2(p)$, the Fourier pairs of $F_1(x)$ and $F_2(x)$. Symbolically:

$$F_1(x) * F_2(x) \rightleftharpoons \Phi_1(p).\Phi_2(p) \qquad (2.5)$$

The applications of this theorem are manifold and profound. Its proof is elementary:

$$C(x) = \int_{-\infty}^{\infty} F_1(x')F_2(x - x')dx'$$

by definition.

Fourier transform both sides (and note that, because the limits are $\pm\infty$, x', is a dummy variable and can be replaced by any other symbol not already in use):

$$\Gamma(p) = \int_{-\infty}^{\infty} C(x)e^{2\pi ipx}dx = \int_{-\infty}^{\infty} \int_{-\infty}^{\infty} F_1(x')F_2(x - x')e^{2\pi ipx}dx'dx \qquad (2.6)$$

Introduce a new variable $y = x - x'$. Then during the x integration x' is held constant and $dx = dy$

$$\Gamma(p) = \int_{-\infty}^{\infty} \int_{-\infty}^{\infty} F_1(x')F_2(y)e^{2\pi ip(x'+y)}dx'dy$$

which can be separated to give:

$$\Gamma(p) = \int_{-\infty}^{\infty} F_1(x')e^{2\pi px'}dx' . \int_{-\infty}^{\infty} F_2(y)e^{2\pi ipy}dy$$

$$= \Phi_1(p).\Phi_2(p)$$

2.3.2 Examples of convolutions

One of the chief uses of convolutions is to generate new functions which are easy to transform using the Convolution Theorem.

2.3.2.1 Convolution of a function with a δ-function, $\delta(x - a)$

$$C(x) = \int_{-\infty}^{\infty} F(x - x')\delta(x' - a)dx' = F(x - a)$$

by the properties of δ-functions. This can be written symbolically as:

$$F(x) * \delta(x - a) = F(x - a)$$

Applying the convolution theorem to this is instructive as it yields the shift theorem:

$$F(x) \rightleftharpoons \Phi(p); \qquad \delta(x - a) \rightleftharpoons e^{-2\pi ipa}$$

so that $F(x - a) = F(x) * \delta(x - a) \rightleftharpoons \Phi(p)e^{-2\pi ipa}$

More interesting is the convolution of a pair of δ-functions with another function:

$$[\delta(x - a) + \delta(x + a)] \rightleftharpoons 2\cos 2\pi pa$$

hence:

$$[\delta(x - a) + \delta(x + a)] * F(x) \rightleftharpoons 2\cos 2\pi pa.\Phi(p) \qquad (2.7)$$

and this is illustrated in figure 2.5. The Fourier transform of a Gaussian $g(x) = e^{-x^2/a^2}$ is, from chapter 1, $a\sqrt{\pi}e^{-\pi^2p^2a^2}$. The convolution of two unequal Gaussian curves, $e^{-x^2/a^2} * e^{-x^2/b^2}$ can then be done, either as a tedious exercise in elementary calculus, or by the convolution theorem:

$$e^{-x^2/a^2} * e^{-x^2/b^2} \rightleftharpoons ab\pi e^{-\pi^2p^2(a^2+b^2)}$$

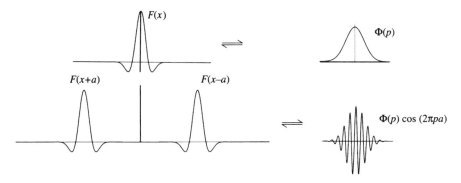

Fig. 2.5. Convolution of a pair of δ-functions with $F(x)$, and its transform.

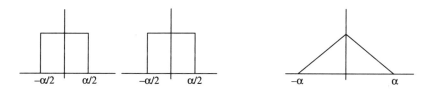

Fig. 2.6. The triangle function, $\Lambda_a(x)$, as the convolution of two top-hat functions.

and the Fourier transform of the right hand side is

$$\frac{ab\sqrt{\pi}}{\sqrt{a^2 + b^2}} e^{-x^2/(a^2+b^2)} \tag{2.8}$$

so that we arrive at a useful practical result:

The convolution of two Gaussians of width parameters a and b is another Gaussian of width parameter $\sqrt{a^2 + b^2}$

or, to put it another way, the resulting half-width is the Pythagorean sum of the two component half-widths.

The convolution of two equal top-hat functions is a good example of the power of the convolution theorem. It can be seen by inspection that the convolution of two top-hat functions, each of height h and width a is going to be a triangle, usually called the 'triangle-function' and denoted by $\Lambda_a(x)$, with height h^2a and base length $2a$.

The Fourier transform of this triangle function can be done by elementary integration, splitting the integral into two parts: $x = -a \rightarrow 0$ and $x = 0 \rightarrow a$. This is tiresome. On the other hand, it is trivial to see that if $h\Pi_a(x) \rightleftharpoons ah\mathrm{sinc}(\pi pa)$ then $h^2a\Lambda_a(x) \rightleftharpoons a^2h^2\mathrm{sinc}^2\pi pa$.

2.3.2.2 The Autocorrelation Theorem

This is superficially similar to the convolution theorem but it has a different physical interpretation. This will be mentioned later in connection with the Wiener–Khinchine Theorem. The autocorrelation function of a function $F(x)$ is defined as:

$$A(x) = \int_{-\infty}^{\infty} F(x')F(x+x')dx'$$

The process of autocorrelation can be thought of as a multiplication of every point of a function by another point at distance x' further on, and then summing all the products: or like a convolution as described earlier, but with identical functions and without taking the mirror-image of one of the two.

There is a theorem similar to the convolution theorem:

$$A(x) = \int_{-\infty}^{\infty} F(x')F(x+x')dx', \text{ by definition.}$$

Fourier transform both sides:

$$\Gamma(p) = \int_{-\infty}^{\infty} A(x)e^{2\pi ipx}dx = \int_{-\infty}^{\infty}\int_{-\infty}^{\infty} F(x')F(x+x')e^{2\pi ipx}dx'dx$$

let $x + x' = y$. Then if x' is held constant, $dx = dy$

$$\Gamma(p) = \int_{-\infty}^{\infty}\int_{-\infty}^{\infty} F(x')F(y)e^{2\pi ip(y-x')dx'}dy$$

which can be separated to

$$\Gamma(p) = \int_{-\infty}^{\infty} F(x)e^{-2\pi ipx'}dx'.\int_{-\infty}^{\infty} F(y)e^{2\pi ipy}dy$$
$$= \Phi^{\star}(p).\Phi(p)$$

so that

$$A(x) \rightleftharpoons \mid \Phi(p) \mid^2$$

The Wiener–Khinchine Theorem, to be described in chapter 4, may be thought of as a physical version of this theorem. It says that if $F(t)$ represents a signal, then its autocorrelation is (apart from a constant of proportionality) the Fourier transform of its power spectrum, $\mid \Phi(v) \mid^2$.

2.4 The algebra of convolutions

You can think of convolution as a mathematical operation analogous to addition, subtraction, multiplication, division, integration and differentiation. There are rules for combining convolution with the other operations. It cannot be associated with multiplication for example, and in general:

$$[A(x) * B(x)].C(x) \neq A(x) * [B(x).C(x)]$$

But convolution signs and multiplication signs can be exchanged across a Fourier transform symbol, and this is very useful in practice. For example:

$$[A(x) * B(x)].[C(x) * D(x)] \rightleftharpoons [a(p).b(p)] * [c(p).d(p)]$$

(obviously upper case and lower case letters have been used to associate Fourier pairs), and as further examples:

$$A(x) * [B(x).C(x)] \rightleftharpoons a(p).[b(p) * c(p)]$$

$$[A(x) + B(x)] * [C(x) + D(x)] \rightleftharpoons [a(p) + b(p)].[c(p) + d(p)]$$

$$[A(x) * B(x) + C(x).D(x)].E(x) \rightleftharpoons [a(p).b(p) + c(p) * d(p)] * e(p)$$

So far as we use Fourier transforms in physics and engineering, we are concerned mostly with functions and manipulations like this to solve problems, and fluency in this relatively easy algebra is the key to success. Computation, rather than calculation is involved, and there is much software available to compute Fourier transforms digitally. However, most computation is done using complex exponentials and these involve the full complex transform. A later chapter deals with this subject.

2.5 Other theorems

2.5.1 The Derivative Theorem

If $\Phi(p)$ and $F(x)$ are a Fourier pair:

$$F(x) \rightleftharpoons \Phi(p), \text{ then } dF/dx \rightleftharpoons -2\pi ip\Phi(p)$$

Proofs are elementary. You can integrate dF/dx by parts or you can differentiate $F(x)$:

$$F(x) = \int_{-\infty}^{\infty} \Phi(p)e^{-2\pi ipx} dp$$

differentiate with repect to x:

$$dF/dx = \int_{-\infty}^{\infty} -2\pi i p \Phi(p) e^{-2\pi i p x} dp$$

$$= -2\pi i \int_{-\infty}^{\infty} p\Phi(p) e^{-2\pi i p x} dp \qquad (2.9)$$

and the right hand side is $-2\pi i$ times the Fourier transform of $p\Phi(p)$.

Example:

The top-hat function $\Pi_a(x) \rightleftharpoons a \operatorname{sinc}\pi p a$. If the top-hat function is differentiated with respect to x, the result is a pair of δ-functions at the points where the slope was infinite:

$$\frac{d\Pi_a(x)}{dx} = \delta(x + a/2) - \delta(x - a/2)$$

Transforming both sides:

$$\delta(x + a/2) - \delta(x - a/2) \rightleftharpoons e^{-\pi i p a} - e^{\pi i p a} = -2i \sin \pi p a$$

$$= -2\pi i p[a \operatorname{sinc}(\pi p a)]$$

The theorem extends to further derivatives:

$$d^n F(x)/dx^n \rightleftharpoons (-2\pi i p)^n \Phi(p)$$

and much use is made of this in mathematics.

Example:

The differential equation of simple harmonic motion is:

$$md^2 F(t)/dt^2 + kF(t) = 0$$

where $F(t)$ is the displacement of the oscillator from equilibrium at time t. If we Fourier-transform this equation, $F(t)$ becomes $\Phi(v)$ and d^2F/dt^2 becomes $-4\pi^2 v^2 \Phi(v)$. The equation then becomes:

$$\Phi(v)(k/m - 4\pi^2 v^2) = 0$$

which, apart from the trivial solution $\Phi(v) = 0$ requires $v = \pm 2\pi \sqrt{k/m}$ – and this is just a small taste of the power which is available for the solution of differential equations using Fourier transforms.

2.5.2 The Convolution Derivative Theorem

$$\frac{d}{dx}[F_1(x) * F_2(x)] = F_1(x) * \frac{dF_2(x)}{dx} = \frac{dF_1(x)}{dx} * F_2(x) \qquad (2.10)$$

The derivative of the convolution of two functions is the convolution of either of the two with the derivative of the other. The proof is simple and is left as an exercise.

2.5.3 Parseval's Theorem

This is met under various guises. It is sometimes called 'Rayleigh's Theorem' or simply the 'Power Theorem'. In general it states:

$$\int_{-\infty}^{\infty} F_1(x)F_2^*(x)dx = \int_{-\infty}^{\infty} \Phi_1(p)\Phi_2^*(p)dp \qquad (2.11)$$

where * denotes a complex conjugate.
The proof of the theorem is in the appendix.

Two special cases of particular interest are :

$$\frac{1}{P}\int_0^P |F(x)|^2 \, dx = \sum_{-\infty}^{\infty}(a_n^2 + b_n^2) = \frac{A_0^2}{4} + \frac{1}{2}\sum_1^{\infty}[A_n^2 + B_n^2] \qquad (2.12)$$

which is used for finding the power in a periodic waveform, and

$$\int_{-\infty}^{\infty} |F(x)|^2 \, dx = \int_{-\infty}^{\infty} |\Phi(p)|^2 \, dp \qquad (2.13)$$

for non-periodic Fourier pairs.

2.5.4 The Sampling Theorem

This is also known as the 'Cardinal Theorem' of interpolary function theory, and originated with Whittaker†, who asked and answered the question: How often must a signal be measured (sampled) in order that all the frequencies present should be detected? The answer is: the sampling interval must be the reciprocal of twice the highest frequency present.

The theorem is best illustrated with a diagram. This highest frequency is sometimes called the 'folding frequency', or alternatively the 'Nyquist' frequency, and is given the symbol v_f.

† J. M. Whittaker, *Interpolary Function Theory*, Cambridge University Press, 1935.

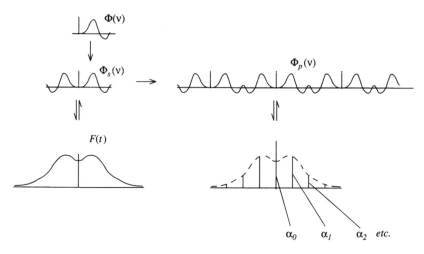

Fig. 2.7. The Sampling Theorem.

Suppose that the frequency spectrum, $\Phi(v)$, of the signal is $F(t)$ and that it contains frequencies from 0 to v_f. From its frequency spectrum $\Phi(v)$ we construct a function $\Phi_s(v)$ which is the original spectrum plus its mirror image. This is symmetrical about the origin and stretches from $-v_f$ to v_f. The convolution of $\Phi_s(v)$ with a Dirac comb of period $2v_f$ provides a periodic function $\Phi_p(v)$ and the Fourier transform of this periodic function is the *product* of a Dirac comb with $\Phi_s(v)$: in other words it is the set of Fourier coefficients representing the series. This is because the Fourier transform of $\Phi_s(v)$ has the same values – apart from a constant – as the coefficients of its periodic counterpart at points $t = 0, 1/2v_f, 2/2v_f, 3/2v_f, \ldots$. Thus the periodic function is always known provided that the coefficients are known, and the coefficients are the values of the original signal $F(t)$, multiplied by a constant, at intervals of time $1/2v_f$. As more coeffcients become known, that is, as more samples are taken, more harmonics can be added to make the spectrum and more detail can be seen in the function when it is reconstructed.

Formally, the process can be written, with $F(t)$ and $\Phi(v)$ a Fourier pair as usual. The Fourier transform of $F(t)III_a(t)$ is:

$$\int_{-\infty}^{\infty} F(t)III_a(t)e^{-2\pi ivt}dt = \Phi(v) * III_{1/a}(v)$$

rewrite the left hand side as:

$$\int_{-\infty}^{\infty} F(t) \sum_{n=-\infty}^{\infty} \delta(t-na)e^{-2\pi i v t} dt = \sum_{n=-\infty}^{\infty} \int_{-\infty}^{\infty} F(t)\delta(t-na)e^{-2\pi i v t} dt$$

$$= \sum_{n=-\infty}^{\infty} F(na)e^{-2\pi i v na} = \Phi'(v)$$

The left hand side is now a Fourier series, so that $\Phi'(v)$ is a periodic function, the convolution of $\Phi(v)$ with a Dirac comb of period $1/a$. The constraint is that $\Phi(v)$ must occupy the interval $-1/2a$ to $1/2a$ only; in other words, $1/a$ is twice the highest frequency in the function $F(t)$, in accordance with the sampling theorem.

2.6 Aliasing

In the sampling theorem it is strictly necessary that the signal should contain no power at frequencies above the folding frequency. If it does, this power will be 'folded' back into the spectrum and will appear to be at a lower frequency. If the frequency is $v_f + v_a$ it will appear to be at $v_f - v_a$ in the spectrum. If it is at twice the folding frequency it will appear to be at zero frequency. For example, a sine-wave sampled at intervals $a, 2\pi + a, 4\pi + a \ldots$ will give a set of samples which are identical. There are, in effect, 'beats' between the frequency and the sampling rate. It is always necessary to take precautions when examining a signal to be sure that a given 'spike' corresponds to the apparent frequency. This can be done either by deliberate filtering of the incoming signal, or by making several measurements at different sampling frequencies. The former is the obvious method but not necessarily the best: if the signal is in the form of a pulse and is in a noisy environment, a lot of the power can be lost by filtering.

Aliasing can be put to good use. If the frequency band stretches from v_0 to v_1 the empty frequency band between v_0 and 0 can be divided into a number of equal frequency intervals each less than $2(v_1 - v_0)$ The sampling interval then need be only $1/2(v_1 - v_0)$ instead of $1/2v_1$. This is a way of demodulating the signal and the spectrum that is recovered occupies the first alias although the original occupied a possibly much higher one. The process is illustrated in figure 2.8.

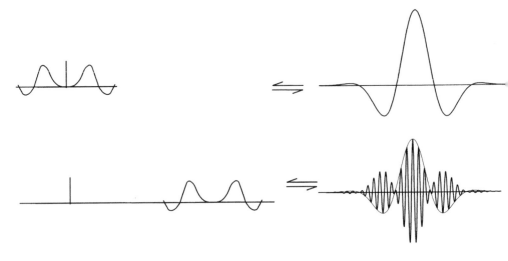

Fig. 2.8. A signal occupying a high alias of a fundamental in frequency space, and its recovery by deliberate undersampling or 'demodulating'.

2.6.1 The Interpolation Theorem

This too comes from Whittaker's interpolary function theory. If the signal samples are recorded, the values of the signal in between the sample points can be calculated. The spectrum of the signal can be regarded as the product of the periodic function with a top-hat function of width $2vf$. In the signal, each sample is replaced by the convolution of the sinc-function with the corresponding δ-function. Each sample, $a_n\delta(t - t_n)$ is replaced by the sinc-function, $a_n \operatorname{sinc}\pi v_f$ and each sinc-function conveniently has zeros at the positions of all the other samples (this is hardly a coincidence, of course) so that the signal can be reconstructed from a knowledge of its samples. which are the coefficients of the Fourier series which form its spectrum.

This is much used in practical physics, when digital recording of data is common, and generally the signal at a point can be well enough recovered by a sum of sinc-functions over twenty or thirty samples on either side. The reason for this is that unless there is a very large amplitude to a sample at some distant point, the sinc-function at a distance of 30π from the sample has fallen to such a low value that it is lost in the noise. It depends obviously on practical details such as the signal/noise ratio in the original data: and more importantly, on the absence of any power at frequencies higher than the folding frequency.

Stated formally, the signal $F(t)$ sampled at times $0, t_0, 2t_0, 3t_0, 4t_0, 5t_0, \ldots$ can be computed at any intermediate point t as the sum

$$F(nt_0 + t) = \sum_{m=-N}^{N} F\{(n+m)t_0\} \operatorname{sinc}\left[\pi(m - t/t_0)\right]$$

where N, infinite in theory, is about $20 \to 30$ in practice. The sum can not be computed accurately near the ends of the data stream and there is a loss of N samples at each end unless fewer samples are taken there.

2.6.2 The Similarity Theorem

This is fairly obvious: if you stretch $F(x)$ so that it is twice as wide, then $\Phi(p)$ will be only half as wide, but twice as high as it was. Formally:

$$\text{if } F(x) \rightleftharpoons \Phi(p) \text{ then } F(ax) \rightleftharpoons |(1/a)| \Phi(p/a)$$

The proof is trivial, and done by substituting $x = ay, dx = ady; p = z/a, dp = (1/a)dz$. Because the integrals are between $-\infty$ and ∞, the variables for integration are 'dummy' and can be replaced by any other symbol not already in use.

2.7 Worked examples

The saw-tooth used in chapter 1 shows an interesting result:

The nth sine-coefficient as we saw, is $-(-1)^n 2h/\pi n$. The sum to infinity of the squares is:

$$\begin{aligned}
\sum_{n=1}^{\infty} \frac{4h^2}{\pi^2 n^2} &= \frac{2}{P} \int_{-P/2}^{P/2} \left[\frac{2hx}{P}\right]^2 dx \\
&= \frac{8h^2}{p^3} \left[\frac{x^3}{3}\right]_{-P/2}^{P/2} \\
&= 2\frac{h^2}{3} = \frac{4h^2}{\pi^2} \sum_{n=1}^{\infty} \frac{1}{n^2}
\end{aligned}$$

so that finally:

$$\sum_{n=1}^{\infty} \frac{1}{n^2} = \frac{\pi^2}{6}$$

This is an example of an arithmetical result coming from a purely analytical calculation. As a way of computing π it is not very efficient:

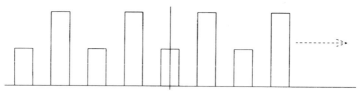

Fig. 2.9.

It is accurate to only six significant figures (3.14159) after one million terms. ($\pi = 6\sin^{-1}(1/2)$, with $\sin^{-1}$ obtained by integrating $1/\sqrt{1-x^2}$ term-by-term, is much more efficient.)

In a rectangular waveform with pulses of length $a/4$ separated by spaces of length $a/4$ and with alternate rectangles twice the height of their neighbours, the amplitude of the second harmonic is greater than the fundamental amplitude.

The waveform can be represented by

$$F(t) = h\Pi_{\frac{a}{4}}(t) * [I\!I\!I_a(t) + I\!I\!I_{\frac{a}{2}}(t)]$$

The Fourier transform is:

$$\Phi(v) = (ah/4)\,\mathrm{sinc}(\pi va/4).[\tfrac{1}{a}I\!I\!I_{\frac{1}{a}}(v) + \tfrac{2}{a}I\!I\!I_{\frac{2}{a}}(v)]$$

and the teeth of this Dirac comb are at $v = 1/a, 2/a\ldots$, with heights

$$(h/4)\,\mathrm{sinc}(\pi/4), (3h/4)\,\mathrm{sinc}(\pi/2), (h/4)\,\mathrm{sinc}(3\pi/4)\ldots,$$

and the ratio of heights of the first and second harmonics is $3/\sqrt{2}$ (figure 2.9).

This effect can be seen in astronomy or radioastronomy when searching for pulsars: the 'interpulses', between the main pulses generate extra power in the second harmonic and can make it larger than the fundamental.

The double-sawtooth waveform: This can not be regarded as the convolution of two rectangular waveforms of equal mark–space† ratio, since the effect of integration is to give an embarrassing infinity. Instead it is the convolution of a top-hat of width a with another identical top-hat

---

† The term 'equal mark–space ratio' comes from radio jargon, and implies that the signal is zero for the same interval that it is not.

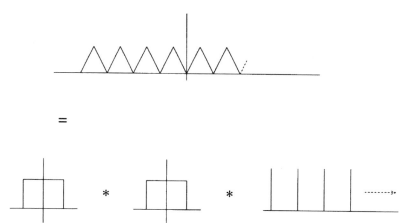

Fig. 2.10. The double-sawtooth waveform.

and with a Dirac comb of period $2a$. Thus:

$$\Pi_a(t) * \Pi_a(t) * III_{2a}(t) \rightleftharpoons (a/2) \operatorname{sinc}^2 \pi v a . III_{\frac{1}{2a}}(v)$$

So that the amplitudes, which occur at $v = 1/2a, 1/a, 3/2a\ldots$ are: $2a/\pi^2, 0, 2(a/3\pi)^2, 0, 2(a/5\pi)^2 \ldots$

3

Applications 1: Fraunhofer diffraction

3.1 Fraunhofer diffraction

The application of Fourier theory to Fraunhofer diffraction problems and to interference phenomena generally, was hardly recognised before the late 1950s. Consequently, only textbooks written since then mention the technique. Diffraction theory, of which interference is only a special case, derives from Huygens' principle: that every point on a wavefront which has come from a source can be regarded as a secondary source: and that all the wavefronts from all these secondary sources combine and interfere to form a new wavefront.

Some precision can be added by using calculus. In figure 3.1, suppose that at O there is a source of 'strength' q, defined by the fact that at A, a distance r from O there is s 'field', E of strength $E = q/r$. Huygens' principle is now as follows:

If we consider an area dS on the surface S we can regard it as a source of strength $E dS$ giving at B, a distance r' from A, a field $E' = qdS/rr'$. All these

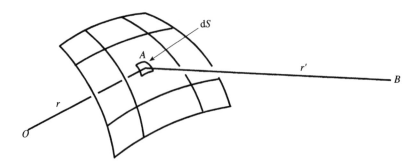

Fig. 3.1. Secondary sources in Fraunhofer diffraction.

40

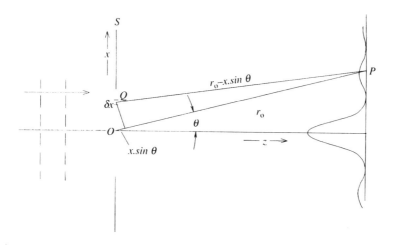

Fig. 3.2. Fraunhofer diffraction by a plane aperture.

elementary fields at *B*, summed over the transparent part of the surface *S*, each with its proper phase†, give the resultant field at *B*. This is quite general – and vague.

In Fraunhofer diffraction we simplify. We assume:

that only two dimensions need be considered. All apertures bounding the transparent part of the surface *S* are rectangular and of length unity perpendicular to the plane of the diagram,

that the dimensions of the aperture are small compared with *r′*,

that *r* is very large so that the field *E* has the same magnitude at all points on the transparent part of *S*, and a slowly varying or constant phase. (Another way of putting it is to say that plane wavefronts arrive at the surface *S* from a source at $-\infty$),

that the aperture *S* lies in a plane.

To begin, suppose that the source, *O* lies on a line perpendicular to the surface *S*, the diffracting aperture. Use Cartesian coordinates, *x* in the plane of *S*, and *z* perpendicular to this (*x* and *z* are traditional here). Then the magnitude of the field *E* at *P* can be calculated.

† Remember: phase change = $(2\pi/\lambda)\times$ path change and the paths from different points on the surface *S* (which, being a wavefront, is a surface of constant phase) to *B* are all different.

Consider an infinitesimal strip at Q, of unit length perpendicular to the (x, z) plane, of width dx and distance x above the z-axis. Let the field strength† there be $E = E_0 e^{2\pi i v t}$. Then the field strength at P from this source will be:

$$d\overline{E}(P) = E_0 dx e^{2\pi i v t} e^{-2\pi i r'/\lambda}$$

where r' is the distance QP. The exponent in this last factor is the *phase difference* between Q and P.

For convenience, choose a time t so that the phase of the wavefront is zero at the plane S, i.e. $t = 0$. Then at P:

$$\overline{E}(P) = \int_{aperture, S} E_0 dx e^{-2\pi i r'/\lambda}$$

and the aperture S may have opaque spots or partially transmitting spots, so that E_0 is generally a function of x.

This is not yet a usable expression.

Now, because $r' \gg x$ (the condition for Fraunhofer diffraction) we can write:

$$r' = r_0 - x \sin \theta$$

and then the field $\overline{E}$ at P is obtained by summing all the infinitesimal contributions from the secondary sources like that at Q, and remembering to include the phase-factor for each. The result is:

$$\overline{E} = E_0 e^{2\pi i r_0/\lambda} \int_{aperture} e^{2\pi i x \sin \theta/\lambda} dx$$

and if we write $\sin \theta / \lambda = p$ we have, finally:

$$\overline{E} = E_0 e^{-2\pi i r_0/\lambda} \int_{-\infty}^{\infty} A(x) e^{2\pi i p x} dx$$

where $A(x)$ is the 'aperture function' which describes the transparent and opaque parts of the screen S. The result of the Fourier transform is to give the *amplitude* diffracted through an angle θ. Where it appears on a screen depends on the distance to the screen, and on whether the screen is perpendicular to the z-direction and other geometrical factors‡.

The important thing to remember is this: that diffraction of a certain

---

† As usual, we use complex variables to represent *real* quantities – in this case the electric field strength. This complex variable is called the 'analytic' signal and the real part of it represents the actual physical quantity at any time at any place.

‡ This is all an approximation: in fact the field *outside* the diffracting aperture is not exactly zero and depends in practice on whether the opaque part of the screen is conducting or insulating. This is a subtlety which can safely be left to post-graduate students.

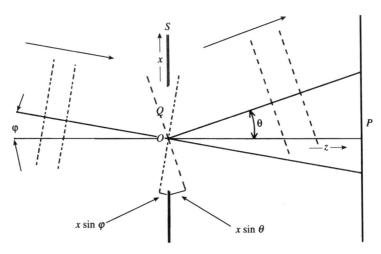

Fig. 3.3. Oblique incidence from a source not on the z-axis.

wavelength at a certain aperture is always *through an angle*: the variable p conjugate to x is $\sin\theta/\lambda$ and it is θ which matters. Diffraction theory alone says nothing about the size of the pattern: that depends on geometry.

Very often, in practice, the diffracting aperture is followed by a lens, and the pattern is observed at the focal plane of this lens. The approximation, that $r' = r_0 - x\sin\theta$ is now exact, since the image of the focal plane, seen from the diffracting aperture, is at infinity.

Problems in Fraunhofer diffraction can thus be reduced to writing down the aperture function, $A(x)$, and taking its Fourier transform. The result gives the amplitude in the diffraction pattern on a screen at a large distance from the aperture. For example, for a simple parallel-sided slit of width a, the aperture function, $A(x)$ is $\Pi_a(x)$. For two parallel-sided slits of width a separated by a distance b between their centres, $A(x)$ is the convolution $\Pi_a(x) * [\delta(x - b/2) + \delta(x + b/2)]$, and so on. Apertures of various sizes are now encompassed by the same formula and the amplitude of the light (or sound, or radio waves or water waves) diffracted by the aperture through an angle θ can be calculated. The *intensity* of the wave is given by the the r.m.s. value of the amplitude $\times$ (complex conjugate) and the factor $e^{2\pi i r_0/\lambda}$ disappears when this is done.

If the original source is not on the z-axis, then the amplitude of E at $z = 0$ contains a phase factor, as in figure 3.3.

$W - W'$ is a wavefront (a surface of constant phase) and if we choose

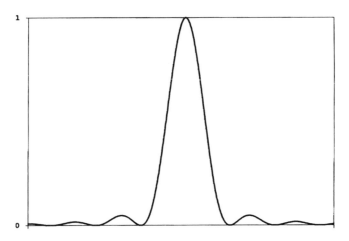

Fig. 3.4. The intensity pattern, $\mathrm{sinc}^2(\pi a \sin\theta/\lambda)$, from diffraction at a single slit.

a moment when the phase is zero at the origin, the phase at x at that moment is given by $(2\pi/\lambda)x.\sin\phi$, and the phase factor that must multiply E_0 is $e^{(-2\pi/\lambda)x\sin\phi}$.

The magnitude at P is then

$$\overline{E} = E_0 e^{2\pi i r_0/\lambda} \int_{-\infty}^{\infty} A(x)e^{(-2\pi i/\lambda)x(\sin\theta+\sin\phi)}dx$$

and when the Fourier transform is done, the oblique incidence is accounted for by remembering that $p = (\sin\theta + \sin\phi)/\lambda$.

3.2 Examples

3.2.1 Single-slit diffraction, normal incidence

For a single slit with parallel sides, of width a, the aperture function is $A(x) = \Pi_a(x)$. Then:

$$\overline{E} = k.\,\mathrm{sinc}(\pi ap) = k.\,\mathrm{sinc}(\pi a \sin\theta/\lambda)$$

(where k is the constant $E_0 a e^{-2\pi i r_0/\lambda}$), and the intensity is this multiplied by its complex conjugate:

$$\overline{EE^*} = I(\theta) = \mid k \mid^2 .\,\mathrm{sinc}^2(\pi a \sin\theta/\lambda) \qquad (3.1)$$

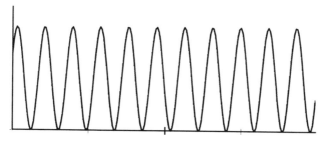

Fig. 3.5. Intensity pattern from interference between two point sources.

3.2.2 Two point sources at $\pm b/2$ (for example, two antennae, transmitting in phase from the same oscillator)

Then:

$$A(x) = \delta(x - b/2) + \delta(x + b/2)$$

and the Fourier transform of this is (chapter 1, equation 1.21):

$$\overline{E} = 2k.\cos(\pi b \sin\theta/\lambda)$$

and the intensity is this amplitude multiplied by its complex conjugate:

$$I(\theta) = 4 \mid k \mid^2 .\cos^2(\pi b \sin\theta/\lambda)$$

3.2.3 Two-slits, each of width a, with centres separated by a distance b (Young's slits, Fresnel's biprism, Lloyd's mirror, Rayleigh's refractometer, Billet's split-lens)

$$A(x) = \Pi_a(x) * [\delta(x - b/2) + \delta(x + b/2)]$$

Then, applying the convolution theorem:

$$I(\theta) = 4k^2 \operatorname{sinc}^2(\pi a \sin\theta/\lambda)\cos^2(\pi b \sin\theta/\lambda)$$

3.2.4 Three parallel slits, each of width a, centres separated by a distance b

To simplify the algebra, put $\sin\theta/\lambda = p$

$$A(x) = \Pi_0(x) * [\delta(x - b) + \delta(x) + \delta(x + b)]$$

$$\begin{aligned} \overline{A}(p) &= k\sin c(\pi pa)[e^{2\pi ibp} + 1 + e^{-\pi ipb}] \\ &= k\sin c(\pi pa)[2\cos(2\pi pb) + 1] \end{aligned}$$

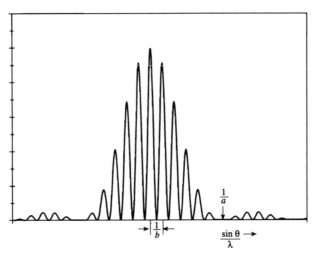

Fig. 3.6. Intensity pattern from interference between two slits of width a separated by a distance b.

and the intensity diffracted at angle θ is:

$$
\begin{aligned}
I(p) &= k^2\mathrm{sinc}^2(\pi pa)[2\cos(4\pi pb) + 4\cos(2\pi pb) + 3] \\
&= k^2\mathrm{sinc}^2(\pi a\sin\theta/\lambda)[2\cos(4\pi b\sin\theta/\lambda) + 4\cos(2\pi b\sin\theta/\lambda) + 3]
\end{aligned}
$$

3.2.5 The transmission diffraction grating

There are two obvious ways of representing the aperture function. In either case we assume that there are N slits, each of width w, each separated from its neighbours by a, the grating constant, and that N is a large $(10^4 \rightarrow 10^5)$ number.

Then, since $A(x) = \Pi_w(x) * \text{Ш}_a(x)$ represents an infinitely wide grating, its width can be restricted by multiplying it by $\Pi_{Na}(x)$, so that the aperture function is:

$$A(x) = \Pi_{Na}(x)[\Pi_w(x) * \text{Ш}_a(x)]$$

So that the diffraction amplitude is:

$$\overline{E}(\theta) = Na.\,\mathrm{sinc}(\pi Na\sin\theta/\lambda) * [\mathrm{sinc}(\pi w\sin\theta/\lambda).\text{Ш}_{(1/a)}(\sin\theta/\lambda)]$$

N.B. the convolution is with respect to $\sin\theta/\lambda$.

A diagram here is helpful: the second factor is a Dirac comb multiplied

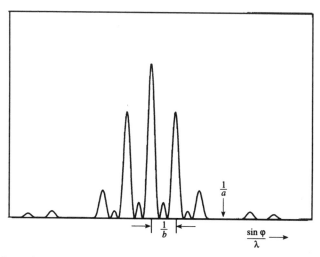

Fig. 3.7. Intensity pattern from interference between three slits of width a, separated by b.

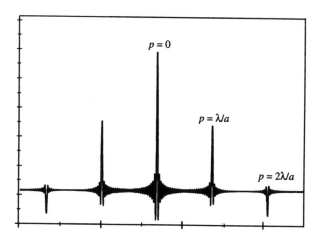

Fig. 3.8. Amplitude transmitted by a diffraction grating.

by a very broad (because w is very small) sinc-function; and the convolution of this with the first factor, a very narrow sinc-function, represents the diffraction produced by the whole aperture of the grating. Since the narrow sinc-function is reduced to insignificance by the time it has reached as far as the next tooth in the Dirac comb, the intensity distribution is this very narrow line profile $\operatorname{sinc}^2(\pi Na \sin\theta/\lambda)$, reproduced at each tooth position with its intensity reduced by the factor $\operatorname{sinc}^2(\pi Wa \sin\theta/\lambda)$.

This is not precise, but is close enough for all practical purposes. To be precise, fastidious and pedantic, the aperture function is

$$A(x) = \sum_{n=0}^{N-1} \delta(x - na) * \Pi_W(x)$$

and since $\delta(x - na) \rightleftharpoons e^{2\pi i n p a}$ the diffracted amplitude is:

$$\overline{E}(\theta) = k\mathrm{sinc}(\pi N W p) \sum_{n=0}^{N-1} e^{2\pi i n p a}$$

where $k = E_0 e^{-2\pi i r_0/\lambda}$ The second factor is a geometrical progression of common ratio $e^{2\pi i p a}$ and after a few lines of algebra the sum is

$$\overline{E}(\theta) = k\mathrm{sinc}(\pi N W p)e^{2\pi i (N-1)pa} \sin(\pi N p a)/\sin \pi p a$$

with $p = \sin\theta/\lambda$ as usual. The intensity distribution is then:

$$I(\theta) = \left(\frac{\sin(\pi N p a)}{\sin(\pi p a)} \right)^2 .\mathrm{sinc}^2(\pi p a) \tag{3.2}$$

In the formula for the intensity, the exponential factor disappears.

If N is large, the first factor is very similar to a sinc-function, especially near the origin, where $\sin \pi p a \simeq \pi p a$ and although it is exact it yields no more information about the diffraction pattern details than the earlier derivation. Either way, the second (sinc^2) factor gives details about the resolution to be obtained, and the first gives information about the angles of diffraction of principal maxima of the diffraction pattern. In particular, if the maximum for one wavelength λ falls at the same diffraction angle θ as the first zero of an adjacent wavelength $\lambda + \delta\lambda$ (the usual criterion for resolution in a grating spectrometer), the two values of p can be compared:

for λ at maximum, $\sin\theta \sim \theta = m\lambda/a$
for λ at first zero, $\theta = m\lambda/a + \lambda/Na$
which is the same angle as for $\lambda + \delta\lambda$ at maximum, i.e. $m(\lambda + \delta\lambda)/a$
whence $\delta\lambda = \lambda/mN$.

which gives the theoretical resolution of the grating.

Two points are worth noting:

(i) No one expects to get the full theoretical resolution from a grating. Manufacturing imperfections reduce it in practice to $\sim 70\%$ of the theoretical value.

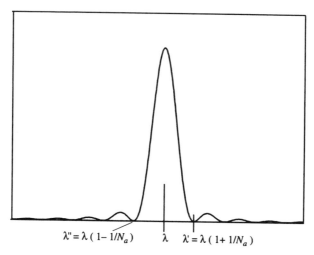

$\lambda'' = \lambda\,(\,1 - 1/N_a\,)$ λ $\lambda' = \lambda\,(\,1 + 1/N_a\,)$

Fig. 3.9. The shape of a spectrum line from a grating. The profile is a sinc^2-function of the form $\text{sinc}^2(\pi Nap)$. The minima at λ' and λ'' are at a wavelength difference $\pm 1/Na$, from the properties of the sinc^2-function.

(ii) Although this is the closest that two wavelengths can still produce separate images, more closely spaced wavelengths can be disentangled if the combined shape is known. The process of *deconvolution* can be used to enhance resolution if need be.

The sinc^2-function in equation 3.1 (illustrated in figure 3.9) represents the intensity near the diffraction image of a monochromatic spectrum line. Although the diffraction intensity defines a direction, θ, in practice a lens or a mirror will focus all the radiation that comes from the grating at the angle θ to a point on its focal surface.

The intensity distribution in the image thus has its width† determined by the width Na of the grating.

Interesting things can be done to the amplitude of the radiation transmitted (or reflected) by the grating by covering the grating with a mask. A diamond-shaped mask for example (figure 3.10) will change the aperture function from $\Pi_a(x)$ to $\Lambda_a(x)$ and the Fourier transform of the aperture function is then:

$$\overline{E}(\theta) = k\,\text{sinc}^2(\pi(aN/2)\sin\theta/\lambda) * [\text{sinc}(\pi w \sin\theta/\lambda).(1/a)III_{(1/a)}(\sin\theta/\lambda)]$$

The shape of the image of a monochromatic line is changed. Instead

---

† By 'width' we mean here the Full Width at Half Maximum Intensity of the spectrum line, usually denoted by 'FWHM'.

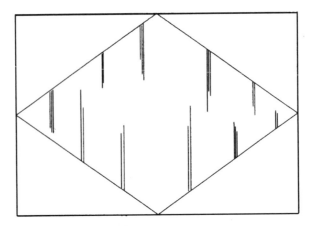

Fig. 3.10. Diffraction grating with a diamond-shaped apodising mask.

of $\text{sinc}^2[\pi Na(\sin\theta/\lambda)]$, it becomes $\text{sinc}^4[(\pi Na/2)(\sin\theta/\lambda)]$. The sinc^4-function is nearly twice as wide as the sinc^2 and the intensity of the light is reduced by a factor of 4, but the intensities of the 'side lobes' are reduced from 1.6×10^{-3} to 2.56×10^{-6} of the main peak intensity. This reduction is important if faint satellite lines are to be identified – for example in studies of fine structure or Raman-scattered lines – where the the satellite intensities are 10^{-6} of the parent or less. The process, which is widely used, in optics and radioastronomy for example, is called *apodising*†. There are more subtle ways of reducing the side-lobe intensities by masking the grating. For example, a mask as in figure 3.11 allows the amplitude transmitted to vary sinusoidally across the aperture according to

$$\Pi_{Na}(x)[A + B\cos(2\pi x/Na)].$$

The Fourier transform of this is

$$\overline{E}(\theta) = \text{sinc}\pi pNa * \{A\delta(p) + B/2(\delta(p - 1/Na) + \delta(p + 1/Na))\}$$

and this is the sum of three sinc-functions, suitably displaced. Figure 3.13 illustrates the effect.

Even more complicated masking is possible and in general what happens is that the power in the side-lobes is redistributed according to the particular problem that is faced. The nearer side-lobes can be suppressed

† From the Greek 'without feet', implying that the side-lobes are reduced or removed.

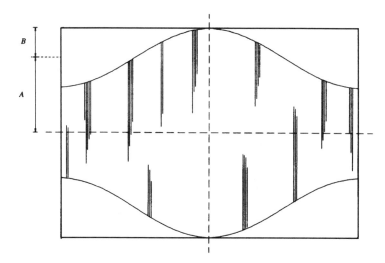

Fig. 3.11. An $A + B\cos(2\pi x/Na)$ apodising mask for a grating.

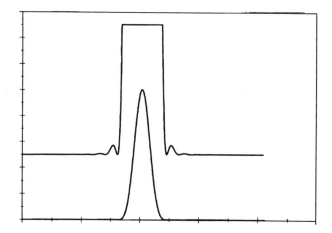

Fig. 3.12. The intensity-profile of a spectrum line from a grating with a sinusoidal apodising mask. The upper curve is the lower curve ×1000 to show the low level of the secondary maxima.

almost completely, for example and the power absorbed into the main peak or pushed out into the 'wings' of the line. Favourite values for A and B are $A = B = 0.5H$ and $A = 0.685H, B = 0.315H$ where H is the length of the grating rulings (*not* the ruled width of the grating).

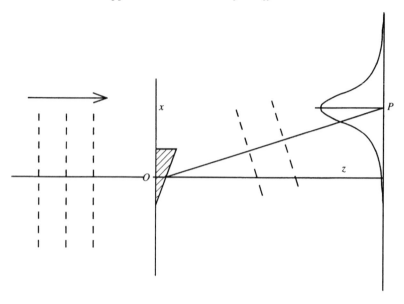

Fig. 3.13. A single-slit aperture with a prism and its displaced diffraction pattern.

3.2.6 *Apertures with phase-changes instead of amplitude changes*

The aperture function may be (indeed *must be*) bounded by a mask edge of finite size and it is possible – for example by introducing refracting elements – to change the phase as a function of x. A prism or lens would do this.

3.2.7 *Diffraction at an aperture with a prism*

Because 'optical' path is $n\times$ geometrical path, the passage of light through a distance x in a medium of refractive index n introduces an *extra* 'path' $(n-1)x$ compared with the same length of path in air or vacuum. Consequently there is a phase change $(2\pi/\lambda)(n-1)x$.

There is thus (figure 3.14) a variation of *phase* instead of transmission across the aperture, so that the aperture function is complex. If the prism angle is ϕ and the aperture width is a, the thickness of the prism at its base is $a\tan\phi$ and when parallel wavefronts coming from $-\infty$ have passed through the prism, the phases at the apex and the base of the prism are 0 and $(2\pi/\lambda)(n-1)a\tan\phi$.

However, we can choose the phase to be zero at the centre of the aperture, and this is usually a good idea because it saves unnecessary algebra later on.

Then the phase at any point x in the aperture is $\zeta(x) = (2\pi/\lambda)x(n-1)\tan\phi$ and the aperture function describing the Huygens wavelets is:

$$A(x) = \Pi_a(x)e^{(2\pi i/\lambda)x(n-1)\tan\phi}$$

The Fourier transform of this, with $p = \sin\theta/\lambda$ as usual, is:

$$\overline{E}(\theta) = A \int_{-a/2}^{a/2} e^{(2\pi i/\lambda)x(n-1)\tan\phi} e^{2\pi ipx} dx$$

so that, after integrating and multiplying the amplitude distribution by its complex conjugate we get:

$$I(\theta) = A^2 a^2 \operatorname{sinc}^2\{a\pi[p + (n-1)\tan\phi/\lambda]\}$$

Notice that if $n = 1$ we have the same expression as in equation 3.1. Here we see that the shape of the diffraction pattern is identical, but that the principal maximum is shifted to the direction $p = \sin\theta/\lambda = -(n-1)\tan\phi/\lambda$ or to the diffraction angle $\theta = \sin^{-1}[(n-1)\tan\phi]$. This is what would be expected from elementary geometrical optics when θ and ϕ are small.

3.2.8 *The blazed diffraction grating*

It is only a small step to the description of the diffraction produced by a grating which comprises, instead of alternating opaque and transparent strips, a grid of parallel prisms. There are two advantages in such a construction. Firstly the aperture is completely transparent and no light is lost, and secondly the prism arrangement means that, for one wavelength at least, all the incident light is diffracted into one order of the spectrum.

The aperture function is, as before, the convolution of the function for a single slit with a Dirac comb, the whole being multiplied by a broad $\Pi_{Na}(x)$ representing the whole width of the grating.

The diffracted intensity is then the same shifted sinc^2-function as above, but multiplied by the convolution of a Dirac comb with a narrow sinc-function, the Fourier pair of $\Pi_{Na}(x)$, which represents the shape of a single spectrum line. Now, there is a difference, because the broad sinc-function produced by a single slit has the same width as the spacing of the teeth in the Dirac comb. The zeros of this broad sinc-function are adjusted accordingly, and for one wavelength, the first order of diffraction falls on its maximum, while all the other orders fall on its zeros. For this wavelength, *all* the transmitted light is diffracted into first order. For adjacent wavelengths the efficiency is similarly high, and in general the efficiency remains usefully high for wavelengths between 2/3 and 3/2 of this wavelength.

This is the 'blaze wavelength' of the grating and the corresponding angle θ is the 'blaze-angle'.

Reflection gratings are made by ruling lines on an aluminium surface with a diamond scribing tip, held at an angle to the surface so as to produce a series of long thin mirrors, one for each ruling. The angle is the 'blaze-angle' that the grating will have, and a similar analysis will show easily that the phase change across one slit is $(2\pi/\lambda)2a\tan\beta$ where β is the 'blaze-angle' and a the width of one ruling (and the separation of adjacent rulings). In practice, gratings are usually used with light incident normally or near-normally on the ruling facets, that is at an incidence angle β to the surface of the grating. There is then a phase change zero across one ruling, but a delay $(2\pi/\lambda)2a\sin\theta$ between reflections from adjacent rulings. If this phase change $= 2\pi$ then there is a principal maximum in the diffraction pattern.

Transmission gratings, universally found in undergraduate teaching laboratories, are usually blazed, and the effect can easily be seen by holding one up to the eye and looking at a fluorescent lamp through it. The diffracted images in various colours are much brighter on one side than on the other.

3.3 Polar diagrams

Since the important feature of Fraunhofer theory is the angle of diffraction, it is sometimes more useful, especially in antenna theory, to draw the intensity pattern with a polar diagram, with intensity as r the length of the radius vector and θ as the azimuth angle. The sinc^2-function then appears as in figure 3.14. Sometimes the logarithm of the intensity is plotted instead, to give the *gain* of the antenna as a function of angle.

A word of caution is appropriate here. Although the basic idea of Fraunhofer diffraction may guide antenna design, and indeed allows proper calculation for so-called 'broadside arrays', there are considerable complications when describing 'end-fire' arrays, or 'Yagi' aerials (the sort used for television reception). The broadside array, which comprises a number of dipoles (each dipole consisting of two rods, lying along

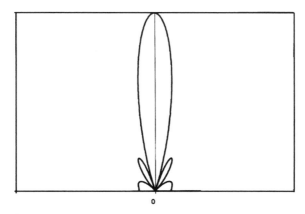

Fig. 3.14. The Polar diagram of a sinc-function.

the same line, each $\lambda/4$ long and with an alternating voltage applied in the middle) behaves like a row of point sources of radiation, and the amplitude at distances large compared with a wavelength can be calculated. Both the amplitude and the relative phase radiated by each dipole can be controlled† so that the shape of the radiation pattern and the strengths of the side-lobes are under control. End-fire antennae, on the other hand, have one dipole driven by an oscillator and rely on resonant oscillation of the other 'passive' dipoles to interfere with the radiation pattern and direct the output power in one direction. The phase re-radiated by a passive dipole depends on whether it is really half a wavelength long, on its conductivity, which is not perfect, and on the dielectric constant of any sheath which may surround it. Consequently, aerial design tends to be based on experience, experiment and computation, rather than on strict Fraunhofer theory. The passive elements may be $\lambda/3$ apart, for example, and their lengths will taper along the direction of the aerial, being slightly shorter on the transmission side and longer on the opposite side to the excited dipole. Such modifications allow a broader band of radiation to be transmitted or received along a narrow cone possibly only a few degrees wide. The nearest optical analogue is probably the Fabry–Perot étalon or, practically the same thing, the interference filter.

---

† Equivalent to apodising in optics, but with more flexibility.

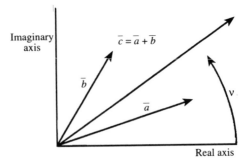

Fig. 3.15. The vector addition of two wave-vectors representing two coherent sources. All three vectors are rotating at the same frequency, v. The vectors are described by the complex numbers $Ae^{2\pi i v t}$, the 'analytic signal', but it is the real part of each – the horizontal component in the graph – which represents the instantaneous value of the electric field of the light wave.

3.4 Phase and coherence

Coherence is an important concept, not only in optics, but whenever oscillators are compared.

No natural light source is exactly monochromatic, and there are small variations in period and hence wavelength from time to time. Two sources are said to be coherent when any small variation in one is matched by a similar variation in the other, so that, for example, if a crest of a wave from one arrives at a given point at the same instant as the trough of a wave from the other, then at all subsequent times troughs and crests will arrive together and there is always destructive interference between the two.

In general two separate sources, two laser beams for example, although nominally of the same wavelength, will not be coherent and no interference pattern will be seen when they both shine on to a screen†. This is why, to generate an interference pattern it is necessary to use two images of the same source as with a Fresnel biprism for example, or two sources fed from the same primary source, as in Young's slits.

The idea can be visualised by thinking of the wave-vector rotating at frequency v. If the source is monochomatic the rotation is exactly at this frequency. Now imagine a rotating coordinate system, rotating at frequency v. The wave-vector will be stationary. A real, physical wave-

---

† This is not strictly true: a very fast detector can 'see' the fringes, which are shifting very rapidly on the surface where they are formed. Exposures in nanoseconds or less are required, and the technology involved is fairly expensive.

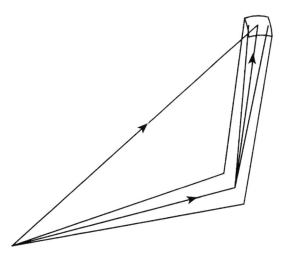

Fig. 3.16. The vector addition of two wave-vectors representing two partially coherent sources. All the vectors are rotating at the same average frequency, but the phase difference ϕ varies randomly over a small range of angles.

vector however will wander about an average direction, and if there are two sources, both vectors will wander independently. If they wander by angles greater than 2π, the vector sum of the two, which represents the resultant amplitude, will vary randomly between $\overline{E_1} + \overline{E_2}$ and $\overline{E_1} - \overline{E_2}$ and the intensity will take an average value $I = I_1 + I_2$, where $I_1 = \,< E_1 E_1^* >$ and the diagonal brackets denote time-averages. On the other hand, if the two sources are coherent the phase angle ϕ between the two vectors will stay constant and the resultant amplitude will be $\overline{E_1} + \overline{E_2}e^{i\phi}$ The intensity of the combined sources will then be

$$I = I_1 + I_2 + 2\sqrt{I_1 I_2}\cos\phi$$

Now a new and useful concept can be introduced: Suppose that the two vectors are not completely independent but that they are loosely coupled together so that the phase-difference ϕ varies about a mean value, and the variation, although random, is through angles less than 2π. The time average of the vector sum is then not simply $I_1 + I_2$, but is less than the vector sum above. Then there will be an interference pattern, but the minima will not be so deep nor the maxima so high as in the fully coherent case. We write:

$$I = I_1 + I_2 + 2\Gamma_{12}\sqrt{I_1 I_2}\cos\phi$$

where ϕ is the average phase-difference. The factor Γ_{12} is always less than unity and is called the *degree of coherence* or the *coherence factor*. The condition is called 'partial coherence'.

It is measured in the laboratory by measuring the maximum and minimum intensities in an interference fringe system. In one case $\phi = 0$ and in the other $\phi = \pi/2$ so that $I_{max} = I_1 + I_2 + 2\Gamma_{12}\sqrt{I_1 I_2}$ and $I_{min} = I_1 + I_2 - 2\Gamma_{12}\sqrt{I_1 I_2}$. The *visibility* of the fringes, which is defined by

$$V = (I_{max} - I_{min})/(I_{max} + I_{min}) = 2\Gamma\sqrt{I_1 I_2}/(I_1 + I_2)$$

is closely related to the degree of coherence. In particular, if the two wave-trains are of equal intensity (and they should be in a well-conducted experiment), then:

$$V = \Gamma_{12}$$

3.5 Exercises

(Note: These examples are not just dry academic solutions of artificial apertures: the results which they provide may well form the physical bases for new types of measuring instrument and servo-control devices.)

Find the angular intensity distribution of the diffracted radiation in the following examples:

(i) An aperture of width A, of which one half has been covered with a transparent strip which delays the wavefront by $\lambda/2$. What happens if the transparent strip slips so as to cover more or less than one half of the aperture? (Method of monitoring and hence controlling the position of the transparent strip.)

(ii) Two apertures, each of width a, with their centres separated by b. One of them is covered by a moving transparent ribbon of varying thickness, causing a varying delay in the wavefront of the order of a few wavelengths on average, with a few tenths of a wavelength variation. (Method of monitoring and hence controlling the thickness during manufacture.)

(iii) Four equi-spaced apertures, the end one and its neighbour covered with a transparent strip of varying thickness. Is there any advantage in using four instead of two? are there optimum values for a and b?

(iv) Two identical half-wave dipole antennae are fed from the same transmitter and one feed incorporates a lossless phase-shifting network. How will the polar diagram of the radiation pattern change as the relative phases of the antennae are changed?

(v) Work out from first principles the theory of the blazed reflection grating. Find the blaze-angle necessary in a reflection grating with 6000 rulings/cm if it is to be perectly efficient in first order for light of wavelength 500 nm (or 5000 Å or 0.5 μm).

4

Applications 2 : Signal analysis and communication theory

4.1 Communication channels

Although the concepts involved in communication theory are general enough to include bush-telegraph drums, alpine yodelling or a ship's semaphore flags, by 'communication channel' is usually meant a single electrical conductor, a waveguide, a fibre-optic cable or a radio-frequency carrier wave. Communication theory covers the same general ground as information theory, which discusses the 'coding' of messages (such as Morse code, not to be confused with encryption, which is what spies do) so that they can be transmitted efficiently. Here we are concerned with the physical transmission by electric currents or radio waves, of the signal or message that has already been encoded. The distinction is that communication is essentially an analogue process, whereas information coding is essentially digital.

For the sake of argument, consider an electrical conductor along which is sent a varying current, sufficient to produce a potential difference $V(t)$ across a terminating impedance of one ohm.

The mean-level or time-average of this potential is denoted by the symbol $< V(t) >$ defined by the equation:

$$< V(t) >= \frac{1}{2T} \int_{-T}^{T} V(t)dt$$

The power delivered by the signal varies from moment to moment, and it too has a mean value:

$$< V^2(t) >= \frac{1}{2T} \int_{-T}^{T} V^2(t)dt$$

For convenience, signals are represented by functions like sinusoids which, in general, disobey one of the Dirichlet conditions described at the beginning of chapter 2: they are not square-integrable:

$$\lim_{t\to\infty} \int_{-T}^{T} V^2(t)dt \to \infty$$

but in practice, the signal begins and ends at finite times and we regard the signal as the product of $V(t)$ with a very broad top-hat function. Its Fourier transform – which tells us about its frequency content – is then the convolution of the true frequency content with a sinc-function so narrow that it can for most purposes be ignored. We thus assume that $V(t) \to 0$ at $|t| > T$ and that

$$\int_{-\infty}^{\infty} V^2(t)dt = \int_{-T}^{T} V^2(t)dt$$

We now define a function $C(v)$ such that $C(v) \rightleftharpoons V(t)$, and Rayleigh's theorem gives:

$$\int_{-\infty}^{\infty} |C(v)|^2 \, dv = \int_{-\infty}^{\infty} V^2(t)dt = \int_{-T}^{T} V^2(t)dt$$

The mean power level in the signal is then:

$$(1/2T) \int_{-T}^{T} |V|^2 (t)dt$$

since $V^2(t)$ is the power delivered into unit impedance; and then:

$$(1/2T) \int_{-T}^{T} |V|^2 (t)dt = \int_{-\infty}^{\infty} \frac{|C(v)|^2}{2T} dv$$

and we *define* $|C(v)|^2 /2T = G(v)$ to be the Spectral Power Density or SPD of the signal.

4.1.1 The Wiener–Khinchine Theorem

The autocorrelation function of $V(t)$ is defined to be:

$$\lim_{t\to\infty} (1/2T) \int_{-T}^{T} V(t)V(t+\tau)dt = \langle V(t)V(t+\tau) \rangle$$

again the integral on the left hand side diverges and we use the shift theorem and the multiplication theorem to give

$$\int_{-T}^{T} V(t)V(t+\tau)dt = \int_{-\infty}^{\infty} C^*(v)C(v)e^{2\pi iv\tau} dv$$

Then:

$$(1/2T) \int_{-T}^{T} V(t)V(t+\tau)dt = \int_{-\infty}^{\infty} \frac{|C(v)|^2}{2T} e^{2\pi i v \tau} dv = R(\tau)$$

so that with the definition of $G(v)$ above:

$$R(\tau) = \int_{-\infty}^{\infty} G(v)e^{2\pi i v \tau} dv$$

and finally:

$$R(\tau) \rightleftharpoons G(v)$$

In other words,

the Spectral Power Density is the Fourier transform of the autocorrelation function of the signal.

This is the Wiener–Khinchine Theorem.

4.2 Noise

The term originally meant the random fluctuation of signal voltage which was heard as a hissing sound in early telephone receivers, and which is still heard in radio receivers which are not tuned to a transmitting frequency. Now it is taken to mean any randomly fluctuating signal which carries no message or 'information'. If it has equal power density at all frequencies it is called 'white' noise†. Its autocorrelation function is always zero since at any time the signal $n(t)$, being random, is as likely to be negative as positive. The only exception is at zero delay, $\tau = 0$, where the integral diverges. The autocorrelation function is therefore a δ-function and its Fourier transform is unity, in accordance with the with the Wiener–Khinchine theorem and with this definition of 'white'.

In practice the band of frequencies which is received is always finite, so that the noise power is always finite. There are other types of noise. For example:

Electron shot noise, or 'Johnson noise', in a resistor, giving a random fluctuation of voltage across it: $< V^2(t) > = 4RkT\Delta v$, where Δv is

---

† This is a rebarbative use of 'white', which really defines a rough surface which reflects all the radiation incident upon it. It is used, less compellingly, to describe the colour of the light emitted by the Sun or even less compellingly, to describe light of constant spectral power density in which all wavelengths (or frequencies: take your choice) contribute equal power.

the bandwidth, R the resistance, k Boltzmann's constant and T the absolute temperature.†

Photon shot noise, which has a normal (Gaussian) distribution of count-rate‡ at frequencies low compared with the average photon arrival rate and, more accurately, a Poisson distribution when equal time-samples are taken. This is met chiefly in optical beams used for communication, and only then when they are weak. Typically, a laser beam delivers 10^{18} photons s^{-1}, so that even at 100 MHz there are 10^{10} photons/sample, or an S/N ratio of $10^5 : 1$.

Semi-conductor noise, which gives a time-varying voltage with a Spectral Power Density which varies as $1/v$ – which is why many semi-conductor detectors of radiation are best operated at high frequency with a 'chopper' to switch the radiation on and off. There is usually an optimum frequency, since the number of photons in a short sample may be small enough to increase photon shot-noise to the level of the semi-conductor noise.

4.3 Filters

By 'filter' we mean an electrical impedance which depends on the frequency of the signal current trying to pass. The exact structure of the filter, the arrangement of resistors, capacitors and inductances, is immaterial. What matters is the effect that the filter has on a signal of fixed frequency and unit amplitude. The filter does two things: it attenuates the amplitude and it shifts the phase. This is all that it does§. The frequency-dependence of its impedance is described by its filter function $Z(v)$. This is defined to be the ratio of the output voltage divided by the input voltage, as a function of frequency:

$$Z(v) = V_o/V_i = A(v)e^{i\phi(v)}$$

where V_i and V_o are 'analytic' representations of the input and output voltages; i.e., they include the phase as well as the amplitude. The impedance is complex since both the amplitude and the phase of V_o may be different from V_i. The filter impedance, Z is usually shown graphically by plotting a polar diagram of the attenuation, A, radially

---

† $V_{\text{rms}} = 1.3 \times 10^{-10}(R\Delta v)^{1/2}$ volts in practice.
‡ Which may be converted into a time-varying voltage by a rate-meter.
§ Unless it is 'active'. Active filters can do other things such as doubling the frequency of the input signal.

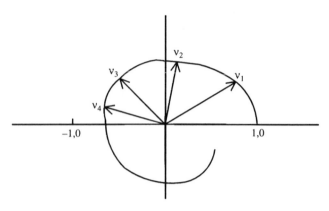

Fig. 4.1. The Nyquist diagram of a typical filter.

against the angle of phase-shift, eliminating v as a variable. The result is called a *Nyquist Diagram*. This is the same figure that is used to describe a feedback loop in servomechanism theory, with the difference that the amplitude A is always less than unity in a passive filter, so that there is no fear of the curve containing the point $(-1, 0)$, the criterion for oscillation in a servomechanism.

4.4 The Matched Filter Theorem

Suppose that a signal $V(t)$ has a frequency spectrum $C(v)$ and a Spectral Power Density $S(v) = |C(v)|^2/2T$. The signal emerging from the filter then has a frequency spectrum $C(v)Z(v)$ and the Spectral Power Density is $G(v)$, given by:

$$G(v) = \frac{|C(v)Z(v)|^2}{2T}$$

If there is white noise passing through the system, with Spectral Power Density $|N(v)|^2/2T$ the total signal power and noise power are:

$$\frac{1}{2T} \int_{-\infty}^{\infty} |C(v)Z(v)|^2 \, dv$$

and

$$\frac{1}{2T} \int_{-\infty}^{\infty} |N(v)Z(v)|^2 \, dv$$

For white noise $|N(v)|^2$ is a constant, $= A$, say, so that the transmitted noise power is:

$$\frac{A}{2T} \int_{-\infty}^{\infty} |Z(v)|^2 \, dv$$

and the ratio of signal power to noise power is the ratio:

$$(S/N)_{power} = \int_{-\infty}^{\infty} |C(v)Z(v)|^2 \, dv / A \int_{-\infty}^{\infty} |Z(v)|^2 \, dv$$

Here we use Schwartz's inequality†

$$\left[\int_{-\infty}^{\infty} |C(v)Z(v)|^2 \, dv \right]^2 \leq \int_{-\infty}^{\infty} |C(v)|^2 \, dv \int_{-\infty}^{\infty} |Z(v)|^2 \, dv$$

so that the S/N power ratio is always $\leq A \int_{-\infty}^{\infty} |C(v)|^2 \, dv$ and the equality sign holds if and only if $C(v)$ is a multiple of $Z(v)$. Hence:

The S/N power ratio will always be greatest if the filter characteristic function $Z(v)$ has the same shape as the frequency content of the signal to be received.

This is the Matched Filter Theorem.

In words, it means that the best signal/noise ratio is obtained if the filter transmission function has the same shape as the signal power spectrum.

It has a surprisingly wide application, in spatial as well as temporal data transmission. The tuned circuit of a radio reciever is an obvious example of a matched filter: it passes only those frequencies containing the information in the program, and rejects the rest of the electromagnetic spectrum. The tone-control knob does the same for the acoustic output. A monochromator does the same thing with light. The 'Radial Velocity Spectrometer' used by astronomers‡ is an example of a spatial matched filter. The negative of a stellar spectrum is placed in the focal plane of a spectrograph, and its position is adjusted sideways – perpendicular to the slit-images – until there is a minimum of total transmitted light. The movement of the mask necessary for this measures the Doppler effect produced by the line-of-sight velocity on the spectrum of a star.

† See, for example, D. C. Champeney, *Fourier Transforms and their Physical Applications* Appendix F, Academic Press, 1973.
‡ Particularly by R. F. Griffin. See *Astrophys. J.* **148**, 465 (1967).

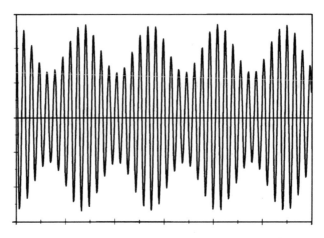

Fig. 4.2. A carrier wave with amplitude modulation.

4.5 Modulations

When a communication channel is a wireless telegraphy channel (a term which comprises everything from a modulated laser beam to an ELF (Extremely Low Frequency) transmitter used to communicate with submerged submarines) it is usual for it to consist of a 'carrier' frequency on which is superimposed a 'modulation'. If there is no modulating signal, the voltage at the receiver varies with time according to:

$$V(t) = V_c e^{2\pi i(v_c t + \phi)}$$

where v_c is the carrier frequency; and the modulation may be carried out by making V, v_c or ϕ a function of time.

4.5.1 Amplitude modulation

If V varies with a modulating frequency v_{mod}, then $V = A + B\cos 2\pi v_{mod} t$ and the resulting frequency distribution will be as in figure 4.2 and as various modulating frequencies from $0 \to v_{max}$ are transmitted, the frequency spectrum will occupy a band of the spectrum from $v_c - v_{max}$ to $v_c + v_{max}$. If low modulating frequencies predominate in the signal, the band of frequencies occupied by the channel will have appearance of figure 4.3 and the filter in the reciever should have this profile too.

The power transmitted by the carrier is wasted unless very low frequencies are present in the signal. The power required from the transmitter can be reduced by filtering its output so that only the range from v_c to

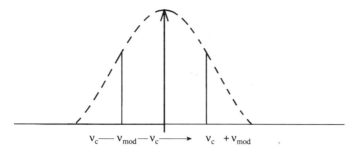

$$v_c - v_{mod} - v_c \longrightarrow \qquad v_c + v_{mod}$$

Fig. 4.3. Various modulating frequencies occupy a band of the spectrum. The time function is $A + B\cos(2\pi v_{mod}t)$ and in frequency space the spectrum becomes the convolution of $V\delta(v - v_c)$ with $A\delta(v) + B(\delta(v - v_0) + \delta(v + v_0))/2$.

v_{max} is transmitted. The receiver is doctored in like fashion. The result is Single Sideband Transmission.

4.5.2 Frequency modulation

This is important because it is possible to increase the bandwidth used by the channel. (By 'channel' is meant here perhaps the radio frequency link used by a spacecraft approaching Neptune and its receiver on Earth, some 4×10^9 Km away.) The signal now is

$$V(t) = A\cos 2\pi v(t)t$$

and $v(t)$ itself is varying according to $v(t) = v_{carrier} + m\cos 2\pi v_{mod}(t)t$. m can be made very large so that for example a voice telephone signal, normally requiring about 3×10^3 Hz bandwidth can be made to occupy several MHz if necessary. The advantage in doing this is found in the Hartley–Shannon Theorem of information theory, which states that the 'channel capacity', the rate at which a channel can transmit information in bits s^{-1} ('bauds') is given by:

$$dB/dt \leq 2\Omega \log_2(1 + S/N)$$

Where Ω is the channel bandwidth, S/N is the *power* signal–noise ratio and dB/dt is the 'baud-rate' or bit-transmission rate.

So, to get a high data transmission rate, you need not slave to improve the S/N-ratio because only the logarithm of that is involved: instead you increase the bandwidth of the transmission. In this way the low power available to the spacecraft transmitter near Neptune is used more effectively than would be possible in an amplitude-modulated transmitter.

Theorems in information theory, like those in thermodynamics, tend to tell you what is possible, without telling you how to do it.

To see how the power is distributed in a frequency-modulated carrier, the message-signal, $a(t)$, can be written in terms of the phase of the carrier signal, bearing in mind that frequency can be defined as rate of change of phase. If the phase is taken to be zero at time $t = 0$, then the phase at time t can be written as:

$$\phi = \int_0^t \frac{\partial \phi}{\partial t} dt$$

and $\partial\phi/\partial t = v_c + \int_0^t a(t)dt$ and the transmitted signal is:

$$V(t) = ae^{2\pi i[v_c + \int_0^t a(t)dt]t}$$

Consider a single modulating frequency v_{mod}, such that $a(t) = k\cos(2\pi v_{mod}t)$. Then

$$2\pi i \int_0^t a(t)dt = \frac{2\pi ik}{2\pi v_{mod}} \sin 2\pi v_{mod}t$$

k is the depth of modulation, and k/v_{mod} is called the *modulation index*, m. Then:

$$V(t) = Ae^{2\pi iv_c t}e^{im\sin(2\pi v_{mod}t)}$$

It is a cardinal rule in applied mathematics, that when you see an exponential function with a sine or cosine in the exponent, there is a Bessel function lurking somewhere. This is no exception. The second factor in the expression for $V(t)$ can be expanded in a series of Bessel functions by the Jacobi expansion†:

$$e^{im\sin(2\pi v_{mod}t)} = \sum_{n=-\infty}^{\infty} J_n(m)e^{2\pi inv_{mod}t}$$

and this is easily Fourier transformable to:

$$\chi(v) = \sum_{n=-\infty}^{\infty} J_n(m)\delta(v - nv_{mod})$$

The spectrum of the transmitted signal is the convolution of $\chi(v)$ with $\delta(v - v_c)$. In other words, $\chi(v)$ is shifted sideways so that the $n = 0$ tooth of the Dirac comb is at $v = v_c$.

† See, for example, Jeffreys & Jeffreys, *Mathematical Physics*, Cambridge University Press, p. 589.

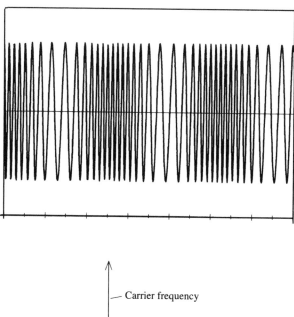

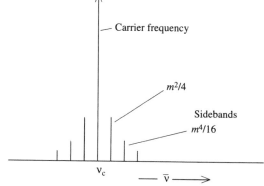

Fig. 4.4. Frequency modulation of the carrier. Many side-bands are present with amplitudes given by the Jacobi expansion.

The amplitudes of the Bessel functions must be computed or looked up in a table† and for small values of the argument are: $J_0(m) = 1$; $J_1(m) = m/2$; $J_2(m) = m^2/4$ etc. Each of these Bessel functions multiplies a corresponding tooth in the Dirac comb of period v_{mod} to give the spectrum of the modulated carrier. Bearing in mind that $m = k/v_{mod}$ we see that the channel is not uniformly filled and there is less power in higher frequencies.

† e.g. Jancke & Emde or Abramowitz & Stegun.

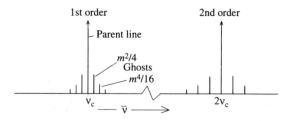

Fig. 4.5. Rowland ghosts in the spectrum produced by a diffraction grating with a periodic error in its rulings.

As an example of the cross-fertilizing effect of Fourier transforms, the theory above can equally be applied to the diffraction produced by a grating in which there is a periodic error in the rulings. In chapter 3 there was an expression for the 'aperture function' of a grating which was

$$A(x) = \Pi_{Na}(x)[\Pi_a(x) * \mathit{III}_a(x)]$$

and if there is a periodic error in the ruling, it is $\mathit{III}_a(x)$ that must be replaced. The rulings, which should have been at $x = 0, a, 2a, 3a \ldots$ will be at $0, a + \alpha \sin(2\pi\beta.a), 2a + \alpha \sin(2\pi\beta.2a), \ldots$etc. and the III-function is replaced by

$$G(x) = \sum_{-\infty}^{\infty} \delta\left[x - na - \alpha \sin(2\pi\beta na)\right]$$

where α is the amplitude of the periodic error, and $1/\beta$ is its 'pitch'. This has a Fourier transform

$$\overline{G}(p) = \sum_{\infty}^{\infty} e^{[na + \alpha \sin(2\pi\beta na)]}$$

with $p = \sin\theta/\lambda$ as in chapter 3. There is a clear analogy with $V(t)$ above. The diffraction pattern then contains what are called 'ghost' lines† around each genuine spectrum line as in figure 4.6.

These satellites which lie on either side of a spectrum line with intensity $\pi^2 p^2 \alpha^2$ times the height of the parent and separated from it by $\Delta\lambda = \pm a\beta\lambda$ are the first order Rowland ghosts. The next ones, of height $\pi^4 p^4 \alpha^4$ of the parent intensity are the second order ghosts, and so on. The analogy with the channel occupation of a frequency-modulated carrier is exact.

† Rowland ghosts, after H. A. Rowland, the inventor of the first effective grating-ruling engine.

There are of course many other ways of modulating a carrier, such as *phase* modulation, *pulse-width* modulation, *pulse-position* modulation, *pulse-height* modulation and so on. Several different kinds of modulation can be applied simultaneously to the same carrier, each requiring a different type of demodulating circuit at the receiver. The design of communications channels includes the art of combining and separating these modulators and ensuring that they do not influence each other with various kinds of 'cross-talk'.

4.6 Multiplex transmission along a channel

There are two ways of sending a number of independent signals along the same communication channel. They are known as *time-multiplexing* and *frequency-multiplexing*. Frequency-multiplexing is the more commonly used. The signals to be sent are used to modulate† a *sub-carrier* which then modulates the main carrier. A filter at the receiving end demodulates the main carrier and transmits only the sub-carrier and its side-bands (which contain the message). Different sub-carriers require different filters and it is usual to leave a small gap in the frequency spectrum between each sub-carrier, to guard against 'cross-talk', that is one signal spreading into the pass-band of another signal.

Time-multiplexing involves the 'sampling' of the carrier at regular time intervals. If, for example, there are ten separate signals to be sent, the sampling rate must be 20 times the highest frequency present in each band. The samples are sent in sequence and switched to ten different channels for decoding, and there must be some way of collating each message channel at the transmitting end with its counterpart at the receiving end so that the right message goes to the right recipient. The 'serial link' between a computer and a peripheral, which uses only one wire, is an example of this, with about eight channels‡, one for each bit-position in each byte of data.

---

† 'Modulate' here means that the main carrier signal is multiplied by the message-bearing sub-carrier. Demodulation is the reverse process, in which the sub-carrier and its message are extracted from the transmitted signal by one of various electronic tricks.

‡ Anywhere between 5 and 11 channels in practice, so long as the transmitter and receiver have agreed beforehand about the number.

4.7 The passage of some signals through simple filters

This is not a comprehensive treatment of the subject, but illustrates the methods used to solve problems. Firstly we need to know about the Heaviside step-function.

4.7.1 The Heaviside step-function

When a switch is closed in an electric circuit there is a virtually instantaneous change of voltage on one side. This can be represented by a 'Heaviside step' function, $H(t)$. It has the property that $H(t) = 0$ for $t < 0$ and $H(t) = 1$ for $t > 0$†. If you differentiate it you get a delta-function $\delta(t)$ and this fact can be used to find its Fourier transform. We use the differential theorem:

$$H(t) = \int_{-\infty}^{\infty} \phi(v)e^{2\pi i v t}dv$$

$$\delta(v) = dH(t)/dt \rightleftharpoons 2\pi i v \phi(v)$$

so that:

$$\phi(v) = 1/2\pi i v$$

4.7.2 The passage of a voltage step through a simple low-pass filter

Suppose that the filter is a 'low-pass' filter with no attenuation or phase-shift up to a critical frequency v_c and zero transmission thereafter. If the height of the step is V volts, the voltage as a function of time is a Heaviside step-function, $VH(t)$. Its frequency content is then $V/2\pi i v$ and the output frequency spectrum is the product of this with the filter profile: that is, $\overline{V}(v) = V/(2\pi i v).\Pi_{v_c}(v))$. The output signal, as a function of time, is the Fourier transform of this, which is

$$f_0(t) = V \int_{-v_c}^{v_c} \frac{e^{2\pi i v t}}{2\pi i v}dv$$

where the top-hat has been replaced by finite limits on the integral.

The function to be transformed is antisymmetric and so there is only a sine-transform:

$$f_0(t) = iV \int_{-v_c}^{v_c} \frac{\sin 2\pi v t}{2\pi i v}dv = Vt \int_{-v_c}^{v_c} \mathrm{sinc}(2\pi v t)dv$$

† Its value *at* $x = 0$ is the subject of debate, but usually taken as $H(0) = 1/2$.

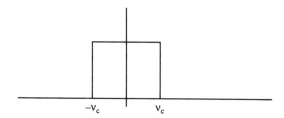

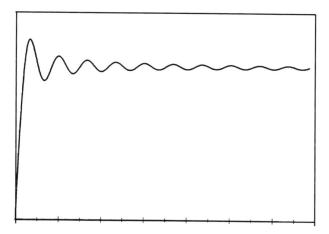

Fig. 4.6. Passage of a Heaviside step-function through a low-pass filter.

$$= 2Vt \int_0^{v_0} \mathrm{sinc}(2\pi vt)dv = \frac{1}{\pi} \int_0^{2\pi v_0 t} \mathrm{sinc}(x)dx$$

with the obvious substitution $x = 2\pi vt$.

The integral is a function of t obviously, and must be computed since sinc-functions are not directly integrable. The result is shown graphically in figure 4.6 The rise-time depends on the filter bandwidth. People who use oscilloscopes on the fastest time-base settings to look at edges will recognise this curve.

4.8 The Gibbs Phenomenon

When you display a square-wave on an oscilloscope, the edges are never quite sharp (unless they are made so by some subtle and deliberate electronic trick) but show small oscillations which increase in amplitude as the corner is approached. They may be quite small in a high-bandwidth oscilloscope.

The reason is found in the finite bandwidth of the oscilloscope. The square-wave is synthesised from an infinite Dirac comb of frequencies, with teeth of heights which depend on the mark–space ratio of the square-wave. To give a *perfect* square-wave, an infinite number of teeth are required, that is to say, the series expansion for $F(t)$ must have an infinite number of terms: sharp corners need high frequencies. Since there is an upper limit to the available frequencies, only a finite number of terms are, in practice, included. This is equivalent to multiplying the Dirac comb in frequency-space by a top-hat function of width $2v_{max}$, and in t-space, which is what the oscilloscope diplays, you see the convolution of the square-wave with a sinc-function $\text{sinc}(2\pi v_{max})$. Convolution with an edge (effectively with a Heaviside step-function) replaces the edge with the integral of the sinc-function between $-\infty$ and t, and the result is shown in figure 4.6.

The phenomenon was discovered experimentally by A.A. Michelson, when he and Stratton designed a mechanical Fourier synthesiser, in which a pen position was controlled by 80 springs pulling together against a master-spring, each controlled by 80 gear-wheels which turned at relative rates of $1/80$, $2/80$, $3/80...79/80$ and $80/80$ turns per turn of a crank-handle. The synthesiser could have the spring tensions set to represent the 80 amplitudes of the Fourier coefficients and the pen position gave the the sum of the series. As the operator turned the crank-handle a strip of paper moved uniformly beneath the pen and the pen drew the graph on it, reproducing, to Michelson's mystification, a square-wave as planned, but showing the Gibbs Phenomenon. Michelson assumed, wrongly, that mechanical shortcomings were the cause: Gibbs gave the true explanation in a letter to *Nature*†.

The machine itself, a marvel of its period, was constructed by Gaertner & Co. of Chicago in 1898. It now languishes in the archives of the South Kensington Science Museum.

4.8.1 The passage of a train of pulses through a low-pass filter

Suppose that we represent the pulse train by a III-function. If the pulse repetition frequency is v_0 the train is described by $III_a(t)$, where $a = 1/v_0$. Suppose that the filter as before, transmits perfectly all frequencies below a certain limit and nothing above that limit. In other words the filter frequency profile or 'filter function' is the same top-hat

† *Nature* **59**, 606, (1899).

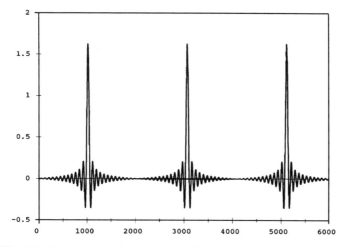

Fig. 4.7. Attenuation of a pulse train by a narrow band low-pass filter.

function Π_{v_f}. The Fourier transforms of the signal and the filter function are $(1/a)III_{v_0}(v)$ and $\Pi_{v_f}(v)$ respectively. The frequency spectrum of the output signal is then the product of the input spectrum and the filter function, $(1/a)III_{v_0}(v).\Pi_{v_f}(v)$ and the output signal is the Fourier transform of this, namely the convolution of the original train of pulses with $\mathrm{sinc}2\pi v_f t$. If the filter bandwidth is wide compared with the pulse repetition frequency, $1/a$, the sinc-function is narrow compared with the separation of individual pulses, and each pulse is replaced, in effect, with this narrow sinc-function. On the other hand if the filter bandwidth is small and contains only a few harmonics of this fundamental frequency, the pulse-train will resemble a sinusoidal wave. An interesting sidelight is that if the transmission function of the filter is a decaying exponential†, $Z(v) = e^{-k|v|}$, then the wave train is the convolution of $III_a(t)$ with $\frac{(k/2\pi^2)}{t^2+(k/2\pi)^2}$. The square of the resulting function may be familiar to students of the Fabry–Perot étalon as the 'Airy' profile.

4.8.2 Passage of a voltage step through a simple high-pass filter

This is an example which shows that contour integration has simple practical uses occasionally:

† Do the Fourier transform of this in two parts: $-\infty \to 0$ and $0 \to \infty$.

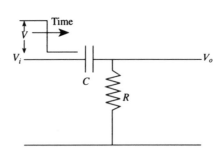

Fig. 4.8. A simple high-pass filter passing a voltage step.

by Ohm's law:

$$V_o = V_i \frac{R}{R + 1/2\pi i v C} = V_i \frac{2\pi i v RC}{2\pi i v RC + 1} = V_i \frac{2\pi i v}{2\pi i v + \alpha}$$

where R is the resistance, C the capacity in the circuit and $\alpha = 1/RC$.

Let the input step have height V so that it is described by the Heaviside step function $\overline{V}_i(t) = VH(t)$. Its frequency content is then $V/2\pi i v = V_i(v)$ and

$$V_o(v) = \frac{V}{2\pi i v} \cdot \frac{2\pi i v}{2\pi i v + \alpha} = \frac{V}{2\pi i v + \alpha}$$

The time-variation of the output voltage is the Fourier transform of this:

$$\overline{V}_o(t) = V \int_{-\infty}^{\infty} \frac{e^{2\pi i v t}}{2\pi i v + \alpha} dv$$

replace $2\pi v$ by z:

$$\overline{V}_o(t) = \frac{V}{2\pi} \int_{-\infty}^{\infty} \frac{e^{izt}}{iz + \alpha} dz$$

multiply top and bottom by $-i$ to clear z of any coefficient:

$$\overline{V}_o(t) = \frac{-iV}{2\pi} \int_{-\infty}^{\infty} \frac{e^{izt}}{z - i\alpha} dz$$

This integral will not yield to elementary methods ('quadrature'). So we use Cauchy's Integral formula†: if z is complex, the integral of $f(z)/(z-a)$ anticlockwise round a closed loop in the Argand plane containing the point a is equal to $2\pi i f(a)$. The quantity $f(a)$ is the *residue* of $f(z)/(z-a)$ at the 'pole', a. Written formally it is:

$$\oint \frac{f(z)}{z - a} dx = 2\pi i f(a)$$

----

† Of fundamental importance and to be found in any book dealing with the functions of a complex variable.

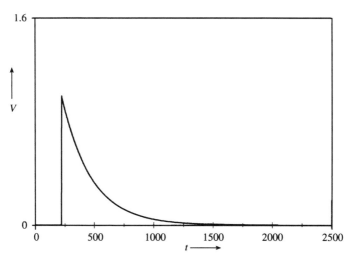

Fig. 4.9. V_o as a function of time simple high-pass filter when the input is a Heaviside step function.

Here the pole is at $z = i\alpha$, so $e^{izt} = e^{-\alpha t}$ and

$$\frac{-iV}{2\pi} \int_C \frac{e^{izt}}{z - i\alpha} dz = -2\pi i \frac{iV}{2\pi} e^{-\alpha t} = V e^{-\alpha t}$$

and the loop ('contour') comprises (a) the real axis, to give the desired integral with $dz = dx$, and (b) the positive semi-circle at infinite radius where the integrand vanishes. Along the real axis the integral is:

$$\lim_{r \to \infty} \frac{-iV}{2\pi} \int_{-r}^{r} \frac{e^{ixt}}{x - i\alpha} dx$$

which is the integral we want. Along the semicircle at large r z is complex and so can be written $z = e^{i\theta}$ or as $r(\cos\theta + i\sin\theta)$ so that e^{izt} becomes $e^{ir(\cos\theta + i\sin\theta)t}$. The real part of this is $e^{-rt\sin\theta}$ which, for positive values of t, vanishes as r tends to infinity (this is why we choose the positive semicircle – $\sin\theta$ is positive). The integral around the positive semicircle then contributes nothing to the total. Thus, for $t > 0$, the time variation $V_o(t)$ of the voltage out, is:

$$V_o(t) = V e^{-\alpha t}$$

For negative values of t, the negative semicircle must be used for integration in order to make the integral vanish. The negative semicircle contains no pole, so the real axis integral is also zero. The complete picture of the response is shown in figure 4.9.

5

Applications 3: Spectroscopy and spectral line shapes

5.1 Interference spectrometry

One of the fundamental formulae of interferometry is the equation giving the condition for maxima and minima in an optical interference pattern:

$$2\mu d \cos \theta = m\lambda$$

and m must be integer for a maximum and half-integer for a minimum.

There are five possible variables in this equation, and by holding three constant, allowing one to be the independent variable and calculating the other, many different types of fringe can be described, sufficient for nearly all interferometers; and nearly all the types of interference fringe refered to in optics texbooks[†], such as 'localised' fringes, fringes of constant inclination, Tolansky fringes, Edser–Butler fringes etc., are included.

5.1.1 The Michelson multiplex spectrometer

Consider the fringes produced by a Michelson interferometer.

If monochromatic light of wavenumber[‡] $\bar{v}$ (= $1/\lambda$) and amplitude A is incident, the beam splitter, if perfect, will send light of amplitude $A/\sqrt{2}$ along each arm. It will be reflected at the two mirrors, and on return to the beam-splitter will recombine with different phases, the result of different path lengths travelled in the two arms. If, for convenience, we choose a moment when the phase is zero at the point of division, the two

[†] e.g. Born & Wolf, *Principles of Optics*, Pergamon Press or Hecht & Zajac, *Optics*, Addison Wesley.
[‡] $\bar{v}$ is used here to denote wavenumber rather than k, since k is sometimes used to mean $2\pi/\lambda$.

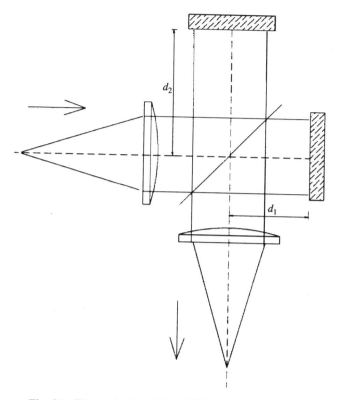

Fig. 5.1. The optical path in a Michelson interferometer.

phases will be $2\pi\bar{v}2d_1$ and $2\pi\bar{v}2d_2$, where the two paths have lengths d_1 and d_2.

The two amplitudes, both complex when the phases are included, are added and the transmitted amplitude is $(A/2)[e^{2\pi i\bar{v}d_1} + e^{2\pi i\bar{v}d_2}]$. The transmitted intensity is then:

$$I_T = (A^2/4)[e^{2\pi i\bar{v}d_1} + e^{2\pi i\bar{v}d_2}][e^{-2\pi i\bar{v}d_1} + e^{-2\pi i\bar{v}d_2}]$$

The path-difference $2(d_1 - d_2)$ is usually written as Δ so that, on completing the multiplication:

$$I_T = (I/2)[1 + \cos(2\pi\bar{v}\Delta)]$$

where $I = AA^*$ is the input intensity.

This describes the fringes which are seen when the path-difference Δ is steadily changed. If, instead of monochromatic radiation of wavenumber

$\bar{v}$, a whole spectrum is used, the intensity at wavenumber $\bar{v}$ will be $I(\bar{v})$. That is to say, the power entering the interferometer between wavenumbers $\bar{v}$ and $\bar{v} + d\bar{v}$ is $I(\bar{v})d\bar{v}$. The intensity emerging will then be

$$dJ(\Delta) = \frac{I(\bar{v})}{2}[1 + \cos(2\pi\bar{v}\Delta)]d\bar{v}$$

and the integral of this over the whole spectrum is:

$$J(\Delta) = \int_{\bar{v}=0}^{\infty} \frac{I(\bar{v})}{2}d\bar{v} + \int_{\bar{v}=0}^{\infty} \frac{I(\bar{v})}{2}\cos(2\pi\bar{v}\Delta)d\bar{v}$$

The first integral is half the total input intensity, I, and we can write $2J(\Delta) - I/2 = K(\Delta)$, so that:

$$K(\Delta) \rightleftharpoons I(\bar{v})$$

Thus if the interferogram $J(\Delta)$ is recorded at suitable intervals of path-difference, the Spectral Power Density $I(\bar{v})$ can be recovered by a Fourier transform. There are some practical difficultes. For example, the path-difference should be increased in *exactly* equal steps, and the intensity emerging from the interferometer should be measured for the same time interval, that is, the same total exposure must be made at each station. As the path difference changes there must be no misalignment of the interferometer mirrors, or the fringe contrast is destroyed. In practice the 'sampling' of the output (the 'interferogram') is never exactly regular. There should be a sample at zero path-difference, and this too is difficult to achieve. The interferogram should be symmetrical about zero path-difference, so that negative path-differences produce the same intensities as the corresponding positive path-differences: usually they do not. These are practical details and they have been overcome, so that Fourier spectroscopy has become a routine technique†.

There are two powerful reasons for doing infra-red spectroscopy this way.

The radiation passing through the interferometer can be received from a large solid angle – hundreds of times larger than in a corresponding grating spectrometer, so that spectra are obtained far more quickly.

There is a so-called 'multiplex advantage'‡, which arises from the fact that radiation from the whole spectral band is received simultaneously by one detector, in contrast to a monochromator, where one wavelength

---

† See, for example, James & Sternberg, *The Design of Optical Spectrometers*, Chapman & Hall, 1969.

‡ Sometimes called the Fellgett Advantage, after its discoverer.

is selected and nearly all the power is discarded inside the instrument. If the spectrum is rich in lines and bands, the signal/noise ratio is increased by a factor in the region of $\sqrt{N}$ where N is the number of resolved elements in the spectrum.

The net result is that, provided the detector is the principal source of noise in the system (which it is in the infra-red, though not in the visible or UV), there is a substantial gain in efficiency: much fainter sources of radiation can be examined, or spectra can be obtained in a much shorter time. For example, the combination of a Fourier multiplex absorption spectrometer with a chromatograph column can be used for on-line analysis of crude oil, where thousands of organic chemical compounds, each with its own characteristic spectrum, pass in sequence through an absorption cell in the spectrometer and can be identified in turn.

The sampling theorem, described in chapter 2, holds: Samples of the interferogram must be taken at intervals Δ_0 of path-difference not greater than† the reciprocal of twice the highest wave number in the spectrum. If necessary the spectrum can be filtered optically to ensure that there is no 'leakage' of higher frequencies into the spectral band. If the spectral band is narrow, the sampling can be at a multiple of the proper interval, so that aliasing can be allowed. A HeNe laser beam can be used to ensure that the samples are taken at the proper intervals. Sometimes, instead of moving a mirror in steps of equal length, stopping, taking a sample, then moving on one step, the path-difference is increased uniformly and smoothly, using the passage of a fringe of laser light (which has a wavelength much shorter than the infra-red spectrum under analysis) to initiate each sample. Then each sample is the *integral* over one step-length of the intensity in the interferogram. What is recorded is the convolution of the interferogram with a top-hat one step-length wide. The spectrum is then product of the true spectrum with a sinc-function with zeros at $\pm 2\bar{\nu}_f$ and the computed spectrum must be divided by this sinc-function.

The other Fourier-related processes discussed earlier also can be applied. A monochromatic line passed through the instrument will yield a sinc-function shape (note: not a sinc^2-function) the result of a finite range of path-differences having been used. This has enormous side-lobes in the modular spectrum with the amplitude of the first side-lobe being 22% of the principal maximum, and apodisation (see chapter 3, page 51) is needed to reduce them. There has been much experimentation with

---

† In practice, usually substantially less than, to leave a gap between the computed spectrum and its mirror image in wavenumber space.

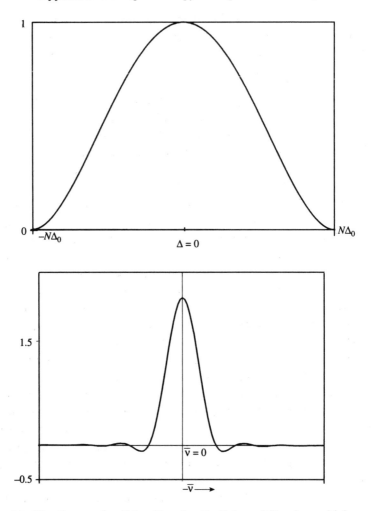

Fig. 5.2. The Connes Apodising Function for Infra-red Fourier multiplex spectroscopy and its ensuing line-profile. Without it the line-profile would be a sinc-function with secondary peaks *below* zero and −22% of the principal maximum in amplitude.

apodising functions – which multiply the interferogram before doing the transform – and a function which multiplies the nth sample of the N-sample interferogram $K(\Delta)$ by $\left[1 - (n/N)^2\right]^2$ due to Janine Connes† has found much favour. It is illustrated in figure 5.2.

---

† J. Connes, *Aspen Conference on Multiplex Fourier Spectroscopy*, 1970, p. 83.

5.2 The shapes of spectrum lines

When an electrical charge is accelerated it loses energy to the radiation field around it. In uniform motion it produces a magnetic field proportional to the current, that is, to $e\partial x/\partial t$; and if the charge is accelerated the changing magnetic field produces an electric field proportional to $e\partial^2 x/\partial t^2$. This in turn induces a magnetic field (via Maxwell's equations) also proportional to $e\partial^2 x/\partial t^2$.

If the charge is oscillating, so are the fields induced around it and these are seen as electromagnetic radiation – in other words, light or radio waves. The power radiated is proportional to the squares of the field strengths $\frac{1}{2}(\epsilon_0 E^2 + \mu_0 H^2)$, which are proportional to $e(\partial^2 x/\partial t^2)^2$. The total power radiated is $2/(3c^2)|\ddot{X}|^2$, where X is the maximum value of the dipole moment ex generated by the oscillating charge. A dipole losing energy in this way is a damped oscillator, and one of Planck's early successes[†] was to show that the damping constant γ is given by:

$$\gamma = \frac{8\pi^2}{3}\frac{c^2}{me}\frac{1}{\lambda^2}$$

The equation of motion for an oscillating dipole is then the usual damped harmonic oscillator equation:

$$\ddot{x} + \gamma\dot{x} + Cx = 0$$

where C is the 'elastic' coefficient, which depends on the particular dipole, and which describes its stiffness and the frequency of the oscillation. γ is of course the damping coefficient which determines the rate of loss of energy.

The solution of the equation is well known, and is:

$$f(t) = e^{-\frac{\gamma}{2}t}\left(Ae^{2\pi i\bar{v}_0 t} + Be^{-2\pi i\bar{v}_0 t}\right)$$

and it is convenient to put $A = 0$ here so that the amplitude, as a function of time is:

$$f(t) = e^{-\frac{\gamma}{2}t}Be^{-2\pi i\bar{v}_0 t}$$

The Fourier transform of this gives the spectral distribution of amplitude and when multiplied by its complex conjugate gives the spectral power density:

$$\phi(\bar{v}) = \int_0^\infty e^{-\frac{\gamma}{2}t}Be^{2\pi i\bar{v}_0 t}e^{-2\pi i\bar{v}t}dt$$

† M. Planck, *Ann. Physik* **60**, 577 (1897).

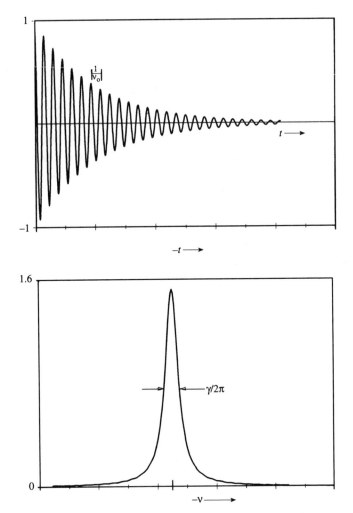

Fig. 5.3. The amplitude of a damped harmonic oscillator and the corresponding spectrum line profile: a Lorentz-function with FWHM $= \gamma/4\pi$. This would be the shape of a spectrum line emitted by an atomic transition if the atoms were held perfectly still while they emitted.

(the lower limit of integration is 0 because the oscillation is deemed to begin then)

$$= e^{-\frac{\gamma}{2}t} \left[\frac{e^{-2\pi i(\bar{v}_0 - \bar{v})t}}{2\pi i(\bar{v}_0 - \bar{v}) - \gamma/2} \right]_0^\infty = \frac{1}{2\pi i(\bar{v}_0 - \bar{v}) - \gamma/2}$$

and the spectral power density is then:

$$I(\bar{v}) = \frac{1}{4\pi^2(\bar{v}_0 - \bar{v})^2 + (\gamma/2)^2}$$

and the line profile is the Lorentz profile discussed in chapter 1.

The same equation can be derived quantum mechanically† for the radiation of an excited atom. The constant $\gamma/2$ is now the 'transition probability', the reciprocal of the 'lifetime of the excited state' if only one downward transition is possible. The FWHM of a spectrum line emitted by an 'allowed' or 'dipole' atomic transition of this sort is usually called the 'natural' width of the line. The shape occurs yet again in nuclear physics, this time called the 'Breit–Wigner formula', and describing in the same way the energy spread in radioactive decay energy spectra. The underlying physics is obviously the same as in the other cases.

There is thus a direct link between the transition probability and the breadth of a spectrum line, and in principle it is possible to measure transition probabilities by measuring this breadth. With typical 'allowed' or 'dipole' transitions – the sort usually seen in spectral discharge lamps – the transition probabilities are in the region of 10^8 s^{-1} and the breadth of a spectrum line at 5000 Å – in the green – is about 0.003 Å. This requires high resolution, a Fabry–Perot étalon for instance, to resolve it. The measurement is quite difficult since atoms in a gas are in violent motion, and a collimated beam of excited atoms is required in order to see the natural decay by this means.

The violent motion of atoms or molecules in a gas is described by the Maxwellian distribution of velocities. The kinetic energy has a Boltzmann distribution, and the fraction of atoms with velocity v in the observer's line-of-sight has a Gaussian distribution:

$$n(v) = n_0 e^{-mv^2/2KT}$$

with a proportionate Doppler shift, giving a Gaussian profile to what otherwise would be a monochromatic line:

$$I(\lambda) = I_0 e^{-(\lambda - \lambda_0)^2/a^2}$$

The width parameter, a, comes from the Maxwell velocity distribution and $a^2 = 2\lambda^2 kT/mc^2$ where k is Boltzmann's constant, T the temperature, m the mass of the emitting species and c the speed of light.

When we substitute numbers in this formula we find that the intensity

---

† See, for example, N.F. Mott & I.N. Sneddon, *Wave Mechanics and its Applications*, Oxford, 1948, Ch 10, §48.

profile is a Gaussian with a FWHM proportional to wavelength, and with $\Delta\lambda/\lambda = 7.16 \times 10^{-7}\sqrt{T/M}$ where M is the molecular weight of the emitting species.

This Doppler broadening, or temperature broadening, by itself would give a different line shape from that caused by radiation damping: a Gaussian profile rather than a Lorentz profile. Unless the emitter has a fairly high atomic weight or the temperature is low, the Doppler width is much greater than the natural width. However, the line shape that is really observed, after making allowance for the instrumental function, is the convolution of the two into what is called a 'Voigt' profile.

$$V(\lambda) = G(\lambda) * L(\lambda)$$

The Fourier transform will be the product of another Gaussian shape and the Fourier transform of the Lorentz shape. This Lorentz shape is a power spectral density and its Fourier transform is, by the Wiener–Khinchine theorem, the autocorrelation of the truncated exponential function representing the decay of the damped oscillator. This autocorrelation is easily calculated. Let s be the variable paired with λ. Then $L(\lambda) \rightleftharpoons l(s)$ where

$$l(s) = \int_{s'}^{\infty} e^{-\frac{\gamma}{2}s}e^{-\frac{\gamma}{2}(s+s')}ds$$

$$= \frac{1}{\gamma}e^{-\frac{\gamma}{2}s} \quad \gamma > 0 \quad ; \quad = \frac{1}{\gamma}e^{\frac{\gamma}{2}s} \quad \gamma < 0$$

Autocorrelations are necessarily symmetrical and so we can write:

$$l(s) = \frac{2}{\gamma}e^{\frac{\gamma}{2}|s|}$$

So long as s is positive, the Fourier transform of the Voigt line profile is the product

$$v(s) = e^{-\pi^2 s^2 a^2} \cdot e^{-\frac{\gamma}{2}s}$$

and a graph of $\log_e v(s)$ versus s is a parabola. From this parabola the two quantities γ and a can be extracted by elementary methods, and the two components of the convolution are separated.

Voigt profiles occur fairly frequently in spectroscopy. Not only is the line-profile of a damped oscillator a Lorentz curve but, the instrumental profile of a Fabry–Perot étalon is the convolution of a Lorentz profile with a Dirac comb. Fabry–Perot fringes, when used to measure the temperature of a gas or a plasma, therefore show Voigt profiles and if the instrument is used properly – that is with the proper spacing between the plates – the Lorentz half-width will be similar to the Gaussian half-width.

Other causes of spectral line shapes can easily be imagined. If the pressure is high, atoms will collide with each other before they have had time to finish their transition. The decaying exponential is then cut short, and the resulting line shape is the convolution of the Lorentz profile with a sinc-function. The width of the sinc-function will be different for every decay, with a Poisson distribution about some average value. The resulting spectrum line then shows 'pressure-broadening', which increases as the intercollision time diminishes, that is, as the pressure increases.

6

Two-dimensional Fourier transforms

6.1 Cartesian coordinates

The extension of the basic ideas to two dimensions is simple and direct. As before, we assume that the function $F(x, y)$ obeys the Dirichlet conditions and we can write:

$$A(p, q) = \int_{y=-\infty}^{\infty} \int_{x=-\infty}^{\infty} F(x, y)e^{2\pi i(px+qy)}dxdy$$

$$F(x, y) = \int_{q=-\infty}^{\infty} \int_{p=-\infty}^{\infty} A(p, q)e^{-2\pi i(px+qy)}dpdq$$

The space of the transformed function is of course two-dimensional, like the original space. The extension to three or more dimensions is obvious.

It sometimes happens that the function $F(x, y)$ is separable into a product $f_1(x)f_2(y)$. In this case the Fourier pair, $A(p, q)$ is separable into $\phi_1(p)\phi_2(q)$ and we find separately that:

$$f_1(x) \rightleftharpoons \phi_1(q); \qquad\qquad f_2(x) \rightleftharpoons \phi_2(q)$$

If $F(x, y)$ is not separable in this way then the transform must be done in two stages:

$$A(p, q) = \int_{-\infty}^{\infty} e^{2\pi iqy} \left\{ \int_{-\infty}^{\infty} F(x, y)e^{2\pi ipx}dx \right\} dy$$

and whether the x-integral or the y-integral is done first may depend on the particular function, F.

6.2 Polar coordinates

Sometimes – often – there is circular symmetry and polar coordinates can be used. The transform space is also defined by polar coordinates,

ρ, ϕ and the substitutions are:

$$x = r \cos \theta; \qquad y = r \sin \theta$$

$$p = \rho \cos \phi; \qquad q = \rho \sin \phi$$

Then

$$A(\rho, \phi) = \int_{r=0}^{\infty} \int_{\theta=0}^{2\pi} F(r, \theta) e^{2\pi i(\rho \cos \phi . r \cos \theta + \rho \sin \phi . r \sin \theta)} r \, dr \, d\theta$$

where $r \, dr \, d\theta$ is now the element of area in the integration, as can be seen directly or from the Jacobean $\partial(x, y)/\partial(r, \theta)$).

This shortens to:

$$A(\rho, \phi) = \int_{r=0}^{\infty} \int_{\theta=0}^{2\pi} F(r, \theta) e^{2\pi i \rho r \cos(\theta - \phi)} r \, dr \, d\theta$$

And if the function F is separable into $P(r)\Theta(\theta)$ the integrals separate into:

$$\int_{r=0}^{\infty} P(r) \left\{ \int_{\theta=0}^{2\pi} \Theta(\theta) e^{2\pi i \rho r \cos(\theta - \phi)} d\theta \right\} r \, dr$$

If there is circular symmetry A is a function of r only. $\Theta(\theta) = 1$ and we write:

$$A(\rho, \phi) = \int_{r=0}^{\infty} P(r) \left[\int_{\theta=0}^{2\pi} e^{2\pi i \rho r \cos(\theta - \phi)} d\theta \right] r \, dr$$

We now put $\theta - \phi = \alpha$, a new independent variable, with $d\alpha = d\phi$ (the integral, being taken around 2π, does not depend on the value of θ).

Then the θ-integral becomes

$$\int_0^{2\pi} e^{2\pi i \rho r \cos \alpha} d\alpha$$

and this (see appendix) is equal to $2\pi J_0(2\pi \rho r)$ where J_0 denotes the zero-order Bessel-function.

Then:

$$A(\rho) = 2\pi \int_0^{\infty} P(r) r J_0(2\pi \rho r) dr$$

which is known as a Hankel transform. It is a close relative of the Fourier transform and, as the Bessel-functions of any order $J_n(x)$ multiplied by $x^{\frac{1}{2}}$, form an orthogonal set† like the trigonometric functions, there is a

---

† If a set is orthogonal and in addition the integral = 1 when the subscripts are the same, the functions are called 'orthonormal'.

similar inversion formula:

$$P(r) = 2\pi \int_0^\infty A(\rho)\rho J_0(2\pi\rho r)d\rho$$

And the two functions are symbolically linked by:

$$P(r) \Leftrightarrow A(\rho)$$

6.3 Theorems

Some, but not all, of the theorems derived in chapter 2 carry over into two dimensions. As above, assume that $P(r) \Leftrightarrow A(\rho)$.

The Similarity Theorem: $P(kr) \Leftrightarrow (1/k^2)A(\rho/k)$

The Addition Theorem: $P_1(r) + P_2(r) \Leftrightarrow A_1(\rho) + A_2(\rho)$

Rayleigh's Theorem:

$$\int_0^\infty \mid P(r) \mid^2 r.dr = \int_0^\infty \mid \Phi(\rho)) \mid^2 \rho.d\rho$$

There is a convolution theorem like that in one dimension but one of the functions has to explore the whole plane in two dimensions instead of just sliding over the other. The product integral is done at each point in the plane to obtain the convolution:

$$C(r^{'}) = P_1(r) ** P_2(r) = \int_{r=0}^\infty \int_{\theta=0}^{2\pi} P_1(r)P_2(R)rdrd\theta$$

where $R^2 = r^2 + r^{'2} - 2rr^{'}\cos\theta$ and the symbol ** is used to denote a two-dimensional convolution.

There is a corresponding convolution theorem:

$$C(r) \Leftrightarrow A_1(\rho)A_2(\rho)$$

6.4 Transforms with circular symmetry

6.4.1 The top-hat function, also known as 'circ' or 'disk'

$$
\begin{aligned}
P(r) &= h, 0 < r < a \\
&= 0, a < r < \infty
\end{aligned}
$$

$$A(\rho) = 2\pi h \int_0^a r.J_0(2\pi\rho r)dr$$

We use

$$\frac{d}{dx}(xJ_1(x)) = xJ_0(x)$$

let $2\pi\rho r = x$; $\qquad 2\pi\rho dr = dx$

$$
\begin{aligned}
A(\rho) &= 2\pi h \int_0^{2\pi a\rho} \frac{x}{2\pi\rho} J_0(x) \frac{dx}{2\pi\rho} \\
&= \frac{h}{2\pi\rho^2} \int_0^{2\pi a\rho} xJ_0(x)dx = \frac{h}{2\pi\rho^2} [xJ_1(x)]_0^{2\pi a\rho} \\
&= \frac{ah}{\rho} J_1(2\pi a\rho) \\
&= \pi a^2 h \left\{ \frac{2J_1(2\pi a\rho)}{2\pi a\rho} \right\}
\end{aligned}
$$

and finally:

$$A(\rho) = \pi a^2 h \operatorname{Jinc}(2\pi a\rho) \text{ where } \operatorname{Jinc}(x) = \frac{2J_1(x)}{x}$$

Jinc contains the factor of 2 in order that $\operatorname{Jinc}(0) = 1$.

This, with appropriate variables r and ρ, gives the amplitude of diffraction of light (or radio waves) at a circular aperure. The intensity distribution, which is the square modulus of this, is the famous 'Airy disc' familiar to students of the telescope and other optical imaging instruments.

6.4.2 The thin annulus

$P(r)$ is a circle of radius a. In optics, a very thin ring transmitting light:

$$P(r) = h\delta(r - a)$$

then:

$$
\begin{aligned}
A(\rho) &= 2\pi h \int_0^\infty r\delta(r - a)J_0(2\pi\rho r)dr \\
&= 2\pi ah J_0(2\pi a\rho)
\end{aligned}
$$

6.5 Applications

The simple two-dimensional Fraunhofer theory of chapter 3 can now be elaborated. There, we assumed that the element dS on the surface S was equal *in area* to dx, the width of a slit $\times$ unit length perpendicular to the diagram.

Now we can use $dS = dxdy$, a small rectangle in the diffracting

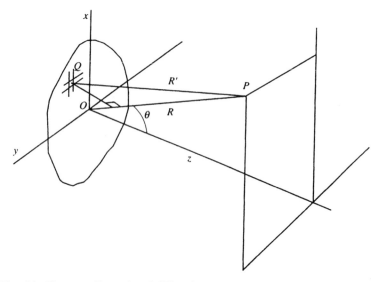

Fig. 6.1. The two-dimensional diffracting aperture, in Cartesian coordinates.

aperture, perpendicular to the direction of progagation, and we can calculate the diffracted amplitude in a direction specified by direction cosines l, m, n . From this we can calculate the intensity at a point on a plane at a distance z from the aperture. If the amplitude at the element of area $dxdy$ at $Q(x, y)$ is $K dxdy$, then at P, on the distant screen, it will be $K dxdy e^{\frac{2\pi}{\lambda} iR'}$ and from elementary coordinate geometry, $R' = R - lx - my$ where l and m are the direction cosines of the line OP and R is the distance from the origin to the point P on the distant screen.

The total disturbance at P is then the sum of all the elementary disturbances from the $z = 0$ plane, so that we can write:

$$A(p, q) = \int\int_{aperture} K dxdy e^{2\pi i(\frac{R}{\lambda} - \frac{lx}{\lambda} - \frac{my}{\lambda})}$$

$$= C \int\int_{aperture} e^{-2\pi i(px+qy)} dxdy$$

where $p = l/\lambda, q = m/\lambda$ and C is a constant which depends on the area of the aperture, and contains the constant phase factors and any other things which do not affect the relative intensity in the diffraction pattern.

If the aperture is a rectangle of side $2a, 2b$ the integrals separate:

$$A(p, q) = C \int_{-a}^{a} e^{-2\pi ipx} dx \int_{-b}^{b} e^{-2\pi iqy} dy$$

and the intensity diffracted in the direction whose direction cosines are $p\lambda, q\lambda$ is the square-modulus of this.

$$I(p,q) = I_0 \operatorname{sinc}^2(2\pi ap) \operatorname{sinc}^2(2\pi bq)$$

Notice that the intensity at the central peak is proportional to the *square* of the area of the aperture.

If the aperture is circular and of radius a, the Hankel transform is used, with $x = r\cos\theta, y = r\sin\theta$ as before and with $p = l/\lambda = \rho\cos\phi$; $q = m/\lambda = \rho\sin\phi$ and $\rho^2 = p^2 + q^2$.

The third direction cosine, n is given by

$$n^2 = 1 - l^2 - m^2 = 1 - (p\lambda)^2 - (q\lambda)^2$$

so that

$$\rho^2 = \frac{1}{\lambda^2}(l^2 + m^2) = \frac{1 - n^2}{\lambda^2}$$

or $\rho = \sin\theta/\lambda$, where θ is the angle between OP and the z-axis.

Then, immediately:

$$A(\theta) = A(0)\frac{J_1(2\pi a\sin\theta/\lambda)}{2\pi a\sin\theta/\lambda}$$

$$I(\theta) = I(0)\left[\frac{J_1(2\pi a\sin\theta/\lambda)}{2\pi a\sin\theta/\lambda}\right]^2$$

Which is the formal equation for the intensity in the Airy disc. Again notice that $I(0)$ is proportional to the square of the area of the aperture. The total power in the pattern is of course proportional to the area of the aperture, but as the radius of the diffracting aperture doubles, for example, the pattern on a distant screen has half the radius and one quarter the area, out to the first zero-intensity ring.

As an exercise, the calculation of the intensity distribution in the diffraction pattern made by an annular aperture can be done. If the inner and outer radii of the annulus are a and b, the amplitude function is:

$$A(\theta) = K\left[a^2\frac{J_1(2\pi a\sin\theta/\lambda)}{2\pi a\sin\theta/\lambda} - b^2\frac{J_1(2\pi b\sin\theta/\lambda)}{2\pi b\sin\theta/\lambda}\right]$$

and the intensity distribution is the square of this.

A graph of this function shows that the central maximum is narrower than that of the Airy disc for the same outer radius. A telescope with an annular aperture apparently beats the 'Rayleigh criterion' for spatial resolution. However, it does so at the expense of putting a lot of intensity into the ring around the central maximum, and the gain is usually more illusory than real.

6.6 Solutions without circular symmetry

In general, provided that the aperture function can be separated into $P(r)$ and $\Theta(\theta)$. Then as we saw earlier:

$$A(\rho, \phi) = \int_{r=0}^{\infty} P(r) \left\{ \int_{\theta=0}^{2\pi} \Theta(\theta) e^{2\pi i \rho r \cos(\theta - \phi)} d\theta \right\} r \, dr$$

Consider the interference pattern of a set of apertures – or antennae – equally spaced around the circumference of a circle. If there are N of them the θ-dependent function is:

$$\Theta(\theta) = \sum_{0}^{N-1} \delta(\theta - 2\pi n/N)$$

and the r-dependent part is

$$P(r) = \delta(r - a)$$

In other words, the sources are equally spaced at angles $2\pi/N$ around the circle of radius a.

Then

$$
\begin{aligned}
A(\rho, \phi) &= \int_0^{\infty} r \delta(r - a) \sum_{0}^{N-1} e^{2\pi i \rho a \cos(2\pi n/N - \phi)} dr \\
&= a \sum_{0}^{N-1} e^{2\pi i \rho a \cos(2\pi n/N - \phi)}
\end{aligned}
$$

This is as far as the analysis can be taken. The pattern, $I(\rho, \phi)$ can be computed without difficulty from this expression, and is a typical example of a problem solved by computer after analysis fails. The particular case of $N = 2$ yields the familiar pattern of two-beam interference, including the hyperbolic shapes of the fringes on a distant plane surface:

$$
\begin{aligned}
A(\rho, \phi) &= a[e^{2\pi i a \rho \cos \phi} + e^{2\pi i a \rho \cos(\pi - \phi)}] \\
&= 2a \cos(2\pi a \rho \cos \phi)
\end{aligned}
$$

and the intensity pattern is given by

$$I(\rho, \phi) = 4a^2 \cos^2(2\pi a \rho \cos \phi)$$

which has maxima when $2a\rho \cos \phi$ is integer. Since $\rho = \sin \alpha/\lambda$ the

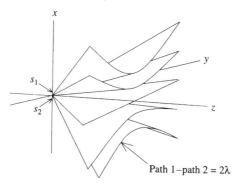

Fig. 6.2. The cones of maximum intensity in a two-beam interference pattern. The two interfering sources are on the x-axis, above and below the origin.

maxima occur when $\phi = n\lambda/2a\sin\alpha$. α is the angle between the z-axis and the direction of diffraction, ϕ is the *azimuth* (angle in the (p, q)-plane, so that interference fringes, the maxima of $I(\rho, \phi)$, emerge along directions defined by the condition $(2a/\lambda)\sin\alpha\cos\phi = $ constant, that is to say, on cones of semi-angle ϕ about the $\phi = 0$-axis. If they are received on a plane perpendicular to the z-axis they show hyperbolic shapes, but on a plane perpendicular to the y-axis (the $\phi = 0$ axis, that is, the axis containing the two sources) the shapes are concentric circles.

Other cases, like $N = 4$ yield to analysis as well. But in general, to parody Clausewitz, it is best to regard computation as the continuation of analysis by other means.

7

Multi-dimensional Fourier transforms

The physical world comprises four dimensions of space and time, and other dimensions, like electrical potential or temperature, are used occasionally, especially for drawing graphs. For this reason Fourier transforms in three or more dimensions can be useful sometimes. The extension is not difficult and can sometimes give greater insight to what is happening in nature than mere geometry.

We use Cartesian coordinates.

7.1 The Dirac wall

This is described by

$$f(x, y) = \delta(x - a)$$

and is zero everywhere except on the line $x = a$, where it is infinite. Despite this infinity, it can be pictured usefully as a wall, parallel to the y-axis, of unit height, as in figure 7.1.

Its two-dimensional Fourier transform is given by:

$$
\begin{aligned}
\phi(p, q) &= \int_{x=-\infty}^{\infty} \int_{y=-\infty}^{\infty} \delta(x - a) e^{2\pi i p x} e^{2\pi i q y} \, dx \, dy \\
&= \int_{y=-\infty}^{\infty} e^{2\pi i p a} e^{2\pi i q y} \, dy \\
&= e^{2\pi i p a} \delta(q)
\end{aligned}
$$

which is complex, but zero except on the line $q = 0$.

Two such Dirac walls, equally spaced about the y-axis have a Fourier transform

$$\phi(p, q) = 2 \cos 2\pi p a . \delta(q)$$

96

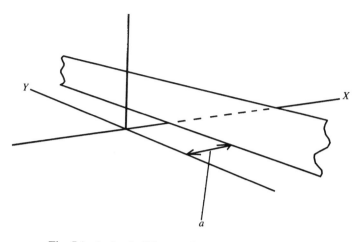

Fig. 7.1. A simple Dirac wall, $f(x, y) = \delta(x - a)$.

A wall inclined to the axes is described by $f(x, y) = \delta(y - mx - c)$ which is zero everywhere except on the line $y = mx + c$, and its Fourier tranform is:

$$\phi(p, q) = \int_{x=-\infty}^{\infty} \int_{y=-\infty}^{\infty} \delta(y - mx - c)e^{2\pi ipx}e^{2\pi iqy} dx dy$$

Do the y-integration first:

$$\phi(p, q) = \int_{x=-\infty}^{\infty} e^{2\pi ipx}e^{2\pi iq(mx+c)} dx$$

Notice that 'integration' here is a simple substitution of the δ-function argument for the variable in the exponential.

$$= e^{2\pi iqc} \int_{x=-\infty}^{\infty} e^{2\pi ix(qm+p)} dx$$
$$= e^{2\pi iqc} \delta(p + qm)$$

and a pair of such walls, equally disposed on either side of the origin has a Fourier transform given by:

$$\delta(y - mx - c) + \delta(y - mx + c) \rightleftharpoons 2\cos 2\pi qc . \delta(p + qm)$$

that is to say, a cosine function, zero everywhere except along the line $p = -qm$.

If we draw the two graphs and superimpose the (p, q) plane on the (x, y) plane, the line of the cosine is perpendicular to the two walls. It

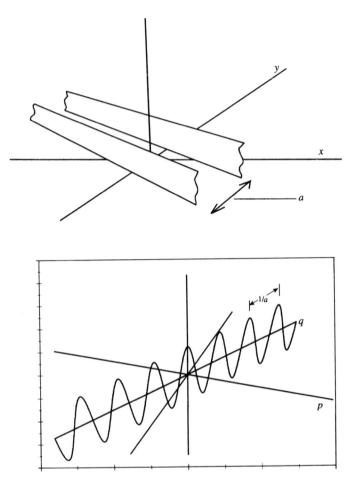

Fig. 7.2. The Fourier transform of a pair of Dirac walls.

is important to notice that the function and its Fourier transform are related in spatial position *irrespective of the orientation of the coordinate system*. It is easy to show, for example, that $\delta(ax - by) \rightleftharpoons \delta(pb + aq)$, walls resting on lines with slopes a/b and $-b/a$ respectively.

7.2 A 'spike' or 'nail'

This is described by a two-dimensional δ-function, $\delta(x - a)\delta(y - b)$ and is zero everywhere except at the point (a, b). As the product of a function of x and a function of y it is separable and its Fourier transform is $e^{2\pi ipa}e^{2\pi iqb}$.

A pair of such nails equally disposed about the origin are described by:

$$f(x, y) = \delta(x - a)\delta(y - b) + \delta(x + a)\delta(y + b)$$

and its Fourier transform is:

$$\phi(p, q) = 2\cos 2\pi(pa + qb)$$

This is a corrugated sheet. Lines of constant phase (wavecrests) lie on the lines $pa + qb$ =integer, and are illustrated in figure 7.3 and again, on superposition, the line joining the nails on the (x, y)-plane is perpendicular to the wavecrests on the (p, q)-plane.

7.3 The Dirac fence

This is an infinite row of equally-spaced δ-functions (the fence posts) along a line. When it runs along the x-axis and the spacing of the posts is a the fence is described by:

$$f(x, y) = \left[\sum_{n=-\infty}^{\infty} \delta(x - na)\right]\delta(y) = III_a(x)\delta(y)$$

Its Fourier transform, $III_{\frac{1}{a}}(p)$ is a parallel set of walls, all parallel to the q-axis and separated by $1/a$.

If the fence is inclined to the x-axis it is described by:

$$f(x, y) = \left[\sum_{n=-\infty}^{\infty} \delta(lx + my - na)\right]\delta(mx - ly)$$

The first factor requires the function to be zero except when $lx + my = na$ (thus defining a set of parallel walls) and the second requires that it be zero except on a line perpendicular to the first set, passing through the origin. This can also be written as:

$$f(x, y) = III_a(lx + my)\delta(mx - ly)$$

where l and m are the direction cosines of the line of the fence and $\sqrt{l^2 + m^2} = 1$.

The Fourier transform can be seen graphically as the convolution of of the two separate transforms. The transform of the first factor is:

$$\phi_1(p, q) = \int_{-\infty}^{\infty}\int_{-\infty}^{\infty}\sum_{n=-\infty}^{\infty}\delta(lx + my - na)e^{2\pi ipx}e^{2\pi iqy}dxdy$$

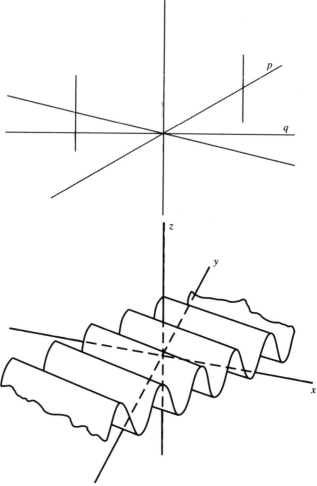

Fig. 7.3. The Fourier transform of a pair of nails at $\pm(x, y)$.

and once more the simple rule for integrating a product which includes a δ-function applies:

$$
\begin{aligned}
\phi_1(p, q) &= \int_{-\infty}^{\infty} \sum_{n=-\infty}^{\infty} e^{2\pi ip(na-my)/l} e^{2\pi iqy} \, dy \\
&= \sum_{n=-\infty}^{\infty} e^{2\pi ipna/l} \int_{-\infty}^{\infty} e^{2\pi iy(q-pm/l)} \, dy \\
&= \sum_{n=-\infty}^{\infty} e^{2\pi ipna/l} \delta(ql - pm)
\end{aligned}
$$

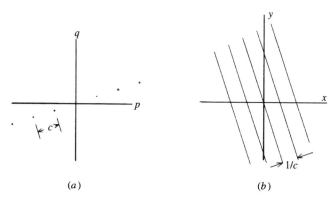

(a)   (b)

Fig. 7.4. A line of fence-posts of spacing c and its Fourier transform, a series of parallel walls a distance $1/c$ apart.

which is a row of fence-posts spaced $1/a$ apart† lying on the line $lq = mp$. The second factor transforms similarly:

$$\phi_2(p,q) = \int_{-\infty}^{\infty} \int_{-\infty}^{\infty} \delta(mx - ly)e^{2\pi ipx}e^{2\pi iqy}dxdy$$

$$= \int_{-\infty}^{\infty} e^{2\pi ip(ly/m)}e^{2\pi iqy}dy/m$$

$$= (1/m)\delta(lp + mq)$$

Which is a wall passing through the origin, lying on the line $lp = -mq$; that is, perpendicular to the fence post of the first factor when the (p,q)-plane is superimposed on the (x, y)-plane.

The convolution of these two factors, $\phi_1(p,q) * *\phi_2(p,q) = w(p,q)$, is an infinite series of parallel walls, spaced $1/a$ apart, lying on lines parallel to the line $lp = -mq$. On superposition of the two spaces, these walls are perpendicular to the original fence line. This is expressed diagrammatically in figure 7.4.

7.4 The 'bed of nails'

Now consider the convolution of two fences, f_1 and f_2. Let each lie on a line through the origin, at angles θ_1 and θ_2 and with spacings a_1 and a_2. The convolution, $f_1 * *f_2$, will be a two-dimensional array of δ-functions – a 'bed of nails'.

† Actually the product of a wall lying on the line $ql = pm$, and an infinite set of walls of spacing a/l lying perpendicular to the p-axis.

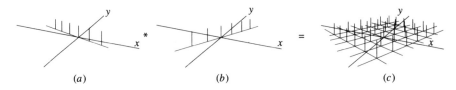

Fig. 7.5. The convolution of two lines of fence-posts to give a 'bed of nails'.

The Fourier transform of this convolution is the product w_1w_2 of the two transforms, each one a series of parallel walls, and differs from zero only when both factors are different from zero. This gives another 'bed of nails'.

The interesting thing is that the route to w_1w_2 from $f_1 * f_2$ is not unique. The two-dimensional array w_1w_2 could have been composed from two different factors, both again parallel sets of walls, but transformed from different fences f_1' and f_2' with different spacings a_1' and a_1' and different angles θ_1' and θ_2'. But the convolution of this new pair will necessarily yield the same function $f_1 * f_2$ as before.

The correspondence between the two beds of nails is this: corresponding to *any* set of parallel lines that can be drawn through points in one plane there is a point† in the other. In figure 7.5, parallel lines separated by $1/a$ in one plane are matched by a point distance a in the other: another set separated by $1/c$ correspond to the point c, and so on. The whole thing is the two-dimensional analogue of the 'reciprocal' lattice idea of the crystallographers.

There is a familiar illustration: seats in a theatre or cinema are arranged regularly, often staggered so that people do not sit directly behind someone. Alignments of seat-backs can be seen in different directions, and these correspond to the lines that can be drawn through beds of nails.

The ideas are even more apparent when transforms are done in three dimensions, when point-arrays are defined by products of three III-functions. $III_a(l_1x + m_1y + n_1z)$ defines a set of parallel planes on which the function is not zero. The planes have equations $l_1x + m_1y + n_1z - \lambda a = 0$ where l, m, n are direction cosines, λ is any integer and a is the perpendicular distance between two adjacent planes.

Two other sets of parallel planes can be defined similarly by

---

† Actually a pair of points – one either side of the origin.

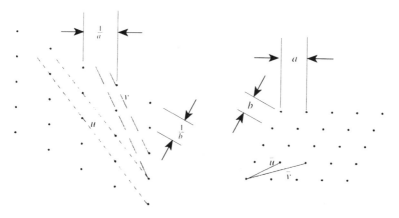

Fig. 7.6. Correspondence between a bed of nails and its Fourier pair.

$III_b(l_2x + m_2y + n_2z)$ and $III_c(l_3x + m_3y + n_3z)$ and the point array or lattice is defined by the product of these three functions.

The Fourier transform of one of these functions is simple:

$$\phi(p,q,r) = \int_{-\infty}^{\infty} \int_{-\infty}^{\infty} \int_{-\infty}^{\infty} \sum_{\lambda=-\infty}^{\infty} \delta(lx + my + nz)e^{2\pi i(px+qy+rz)}dxdydz$$

do the x-integral first:

$$\phi(p,q,r) = \sum_{\lambda=-\infty}^{\infty} \int_{-\infty}^{\infty} \int_{-\infty}^{\infty} e^{2\pi ip(\lambda a-nz-my)/l}e^{2\pi i(qy+rz)}dydz$$

where the integral as before is merely the substitution of the value in the δ-function argument which makes it non-zero.

This is now separable:

$$\begin{aligned}
\phi(p,q,r) &= \sum_{\lambda=-\infty}^{\infty} e^{2\pi ipa\lambda/l} \int_{-\infty}^{\infty} e^{-2\pi i(\frac{pn}{l}-r)z}dz \int_{-\infty}^{\infty} e^{-2\pi i(\frac{pm}{l}-q)y}dy \\
&= \sum_{\lambda=-\infty}^{\infty} e^{2\pi ipa\lambda/l}.\delta\left(\frac{p}{l} - \frac{r}{n}\right)\delta\left(\frac{p}{l} - \frac{q}{n}\right)
\end{aligned}$$

The last two factors, the δ-functions, define two planes. The intersection of the planes defines a line. The λ-sum defines those points on the line where the lattice points exist.

Again if the (p,q,r)-space is superimposed on the (x,y,z)-space, we find that $\phi(p,q,r)$ is a set of equispaced points along a line perpendicular to the set of planes defined by $\delta(lx + my + nz - \lambda a)$ and that the spacing between the points is $1/a$.

A complete three-dimensional lattice, described by the product of three *III*-functions has as its Fourier transform the triple convolution of three lines of equispaced points. This gives a new lattice – the *reciprocal lattice* of the crystallographers. Points on this lattice define the various planes which can be envisaged, containing two-dimensional arrays of lattice points. Lines to these points from the origin define both in length and direction the orientation and separation of lattice planes in (x, y, z)-space.

This now clears up a fundamental problem in describing crystals. The three *III*-functions used to define the crystal lattice in (x, y, z)-space are not the only possible ones. Other sets of three planes can be used – an infinite number of possibilities exists. The points in the reciprocal lattice define such sets of parallel planes, both in direction and separation. Lines ('vectors') from the origin to these points in (p, q, r)-space are normal to the lattice-planes in (x, y, z)-space. The length of each vector is inversely proportional to the separation of the planes in (x, y, z)-space. The coordinates of the lattice points in (p, q, r)-space, when multiplied by a factor to make them integer, are the *Miller Indices* of the (x, y, z)-planes.

8

The formal complex Fourier transform

In physics we are usually concerned with functions of real variables, often experimental curves, data strings, or shapes and patterns. Generally the function is asymmetric about the y-axis and so its Fourier transform is a complex function of a real variable; that is, for any value of p, a complex number is defined.

Any function obeying the Dirichlet conditions can be divided into a symmetric and an antisymmetric part. In figure 8.1, for example, and generally, $f_s(x) = \frac{1}{2}[f(x) + f(-x)]$ and $f_a(x) = \frac{1}{2}[f(x) - f(-x)]$. The symmetric part is synthesised only from cosines and the antisymmetric part only from sines. We write:

$$f(x) = f_s(x) + f_a(x); \quad f_s(x) \rightleftharpoons \phi_s(p); \quad f_a(x) \rightleftharpoons \phi_a(p)$$

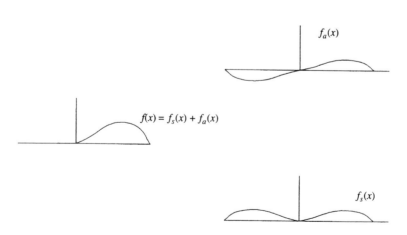

Fig. 8.1. Dividing a function into symmetric and antisymmetric parts.

105

where $\phi_s(p)$, being made of cosines, is real and symmetric and $\phi_a(p)$ is imaginary and, being made of sines, is antisymmetric.

We can also define:

(i) The phase transform of $f(x)$ which is the function $\theta(p)$ where

$$\tan \theta(p) = \phi_a(p)/\phi_s(p)$$

(ii) The power transform

$$P(p) = \mid \phi(p) \mid^2 = \phi_a(p)^2 + \phi_s(p)^2$$

(iii) The modular transform

$$M(p) = \mid \phi(p) \mid = \sqrt{\phi_a(p)^2 + \phi_s(p)^2}$$

All these have their practical uses although none of them has an unique inverse.

A useful corollary of the convolution theorem is that if $C(x) = f_1(x) * f_2(x)$ and if $C(x) \rightleftharpoons \Gamma(p)$ then the power transforms of C, f_1 and f_2, given by $\mid \Gamma \mid^2$, $\mid \phi_1 \mid^2$ and $\mid \phi_2 \mid^2$ are related by:

$$\mid \Gamma \mid^2 = \mid \phi_1 \mid^2 \cdot \mid \phi_2 \mid^2$$

A simple example shows the use of phase transforms. Consider for instance, a displaced top-hat function (any function would do, in fact), of width a and displaced sideways a distance b.

The function is:

$$f(x) = \Pi_a(x) * \delta(x - b)$$

Its Fourier transform is

$$
\begin{aligned}
\phi(p) &= a \operatorname{sinc}(\pi ap).e^{2\pi ibp} \\
&= a \operatorname{sinc}(\pi ap) [\cos(2\pi bp) + i \sin(2\pi bp)]
\end{aligned}
$$

and its phase transform is

$$\theta(p) = \tan^{-1}(\sin 2\pi pb / \cos 2\pi pb) = 0, 2\pi pb \ldots$$

so that $\theta(p) = 0$ when $p = 1/b$.

Phase transforms are useful when an experimentally measured function, which should have been symmetrical, has been displaced by an unknown amount from its axis of symmetry – for example by sampling it in the wrong places. A quick calculation of a few points on the phase transform

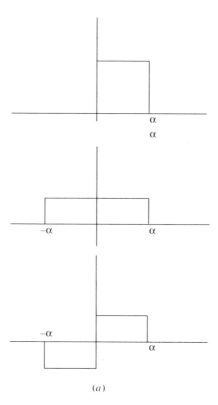

(a)

Fig. 8.2. A top-hat function displaced by its own width. (a) The dissection of the top-hat into symmetric and antisymmetric parts.

will find the displacement and allow any adjustments to be made or the true, symmetrical samples to be computed by interpolation. It also confirms (or not!) that the function really is symmetric, since only then is its phase transform a straight line.

It is worth including here something which will be useful later when considering computing Fourier transforms. Because it is easy to separate the real and imaginary parts of a complex function of x or p, and then to divide these into their symmetric and antisymmetric parts, it is possible to combine two real functions of x into a complex function and then separate the combined complex Fourier transform into its constituent parts. This is a useful technique when computing digital Fourier transforms: one can do two transforms for the price of one. Written analytically:

Let the two functions be $f_1(x)$ and $f_2(x)$ and separate each into its

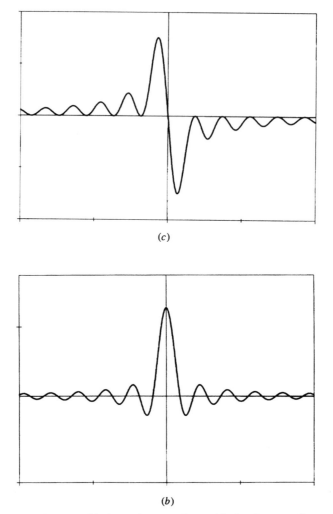

(c)

(b)

Fig. 8.2 *cont.* (b) the cosine transform, (c) the sine transform.

symmetric and antisymmetric parts:

$$f_1(x) = f_{1s}(x) + f_{1a}(x); \quad f_2(x) = f_{2s}(x) + f_{2a}(x)$$

$$\text{let } F(x) = f_1(x) + if_2(x)$$

$$\text{let } F(x) \rightleftharpoons \Phi(p). \text{ Then } \Phi(p) = \int_{-\infty}^{\infty} [f(_1(x) + if_2(x)]e^{2\pi i p x}dx$$

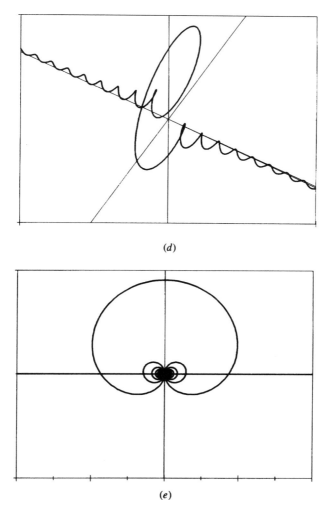

(*d*)

(*e*)

Fig. 8.2 *cont.* (d) the transform in perspective, (e) the Nyquist diagram – the view looking along the *v*-axis.

and remember that a symmetric function has only a cosine transform, etc.

$$\Phi(p) = \int f_{1s}(x)\cos 2\pi px\,dx + i\int f_{1a}(x)\sin 2\pi px\,dx$$
$$+i\int f_{2s}(x)\cos 2\pi px\,dx - \int f_{2a}(x)\sin 2\pi px\,dx$$
$$= \phi_{1s}(p) + i\phi_{1a}(p) + i\phi_{2s}(p) - \phi_{2a}(p)$$

where the meaning of the suffixes is the same as in the f-functions.

Then:

$$\Phi(p) = [\phi_{1s}(p) - \phi_{2a}(p)] + i[\phi_{1a}(p) + \phi_{2s}(p)]$$

Both the real and imaginary parts of $\Phi(p)$ now have symmetrical and antisymmetrical components. When $\Phi(p)$ has been computed, it has a real part, $\Phi_r(p)$ and an imaginary part $\Phi_i(p)$.

The symmetric real part is $\qquad \frac{1}{2}[\Phi_r(p) + \Phi_r(-p)] = \phi_{1s}(p)$

and the antisymmetric part is $\qquad \frac{1}{2}[\Phi_r(p) - \Phi_r(-p)] = -\phi_{2a}(p)$.

Similarly, $\qquad\qquad\qquad\quad \frac{1}{2}[\Phi_i(p) - \Phi_i(-p)] = \phi_{1a}(p)$

and $\qquad\qquad\qquad\qquad\; \frac{1}{2}[\Phi_i(p) + \Phi_i(-p)] = \phi_{2s}(p)$.

so that, finally:

$$f_1(x) \;\rightleftharpoons\; \frac{1}{2}[\Phi_r(p) + \Phi_r(-p)] + \left(\frac{i}{2}\right)[\Phi_i(p) - \Phi_i(-p)]$$

$$\rightleftharpoons\; \frac{1}{2}\phi_{1s}(p) + \left(\frac{i}{2}\right)\phi_{1a}(p)$$

and similarly,

$$f_2(x) \rightleftharpoons \frac{1}{2}\phi_{2s}(p) + \left(\frac{i}{2}\right)\phi_{2a}(p)$$

In other words, the FT of $f_1(x)$ is $\frac{1}{2} \times$ (the symmetrical part of the real component of $\Phi(p)$ plus $i\times$ the antisymmetrical part of the imaginary component of $\Phi(p)$), and the FT of $f_2(x)$ is $\frac{1}{2} \times$ (the symmetrical part of the imaginary component $+ i\times$ the antisymmetric part of the real component). The computer sorts these out without difficulty!

Notice that all the F's, Φ's, f's and ϕ's with suffixes are *real* quantities. This is because a computer deals ultimately in real numbers, although its program may include complex arithmetic. This level of complication is not commonly met when discussing analytic Fourier transforms. However computing algorithms perform the complex transform whether you like it or not, and the relations above can be used to do tricks which shorten computing time when you know that the data represent only real functions.

Diagrammatically, the process can be represented by:

$$f_{1s}(x) \leftarrow \cos \rightarrow \phi_{1s}(p)$$

$$f_{1a} \leftarrow i\sin \rightarrow i\phi_{1a}$$

$$if_{2s} \leftarrow \cos \rightarrow i\phi_{2s}$$

$$if_{2a} \leftarrow i \sin \rightarrow -\phi_{2a}$$

A function is said to be Hermitian if its real part is symmetric and its imaginary part is antisymmetric. So if $f_1(x)$ is symmetric and $f_2(x)$ is antisymmetric, then $\phi_{1a} \equiv 0$ and $\phi_{2s} \equiv 0$. Then

$$\Phi(p) = \phi_{1s}(p) + \phi_{2a}(p)$$

and is real. Alternatively, the Fourier transform of a real but asymmetric function is Hermitian:

$$f_1(x) \rightleftharpoons \phi_{1s}(p) + i\phi_{1a}(p)$$

9

Discrete and digital Fourier transforms

9.1 History

Fourier transformation is formally an analytical process which uses integral calculus. In experimental physics and engineering, however, the integrand may be a set of experimental data, and the integration is necessarily done artificially. Since a separate integration is needed to give each point of the transformed function, the process would become exceedingly tedious if it were to be attempted manually and many ingenious devices have been invented for performing Fourier transforms mechanically, electrically, acoustically and optically. These are all now part of history since the arrival of the digital computer and more particularly since the discovery – or invention – of the so-called 'Fast Fourier Transform' algorithm or FFT as it is generally called. Using this algorithm, the data are put ('read') into a file (or 'array', depending on the computer jargon in use); the transform is carried out, and the file then contains the points of the transformed function. It can be achieved by a software program, or by a purpose-built integrated circuit. It can be done very quickly so that vibration-sensitive instruments with Fourier transformers attached can be used for tuning pianos and motor engines, for aircraft and submarine detection and so on. It must not be forgotten that the ear is Nature's own Fourier transformer†, and, as used by an expert piano-tuner for example, is probably the equal of any electronic simulator in the 20–20 000 Hz range. The diffraction grating too, is a passive Fourier transformer device provided that it is used as a spectrograph taking full advantage of the simultaneity of outputs.

The history of the FFT is complicated and has been researched by Brigham‡ and, as with many discoveries and inventions, it arrived before

† It detects the *power* transform, and is not sensitive to phase.
‡ E. Oran Brigham, *The Fast Fourier Transform*. Prentice-Hall, 1974.

the (computer) world was ready for it. Its digital apotheosis came with the publication of the 'Cooley–Tukey' algorithm[†] in 1965. Since then other methods have been virtually abandoned except for certain specialized cases and this chapter is a description of the principles underlying the FFT and how to use it in practice.

9.2 The discrete Fourier transform

There is a pair of formulae by which sets of numbers $[a_n]$ and $[A_m]$, each set having N elements, can be mutually transformed:

$$A(m) = \frac{1}{N} \sum_0^{N-1} a(n)e^{2\pi inm/N} \quad ; \quad a(n) = \sum_0^{N-1} A(m)e^{-2\pi inm/N} \qquad (9.1)$$

In appearance and indeed in function, these are very similar to the formulae of the analytic Fourier transform and are generally known as a 'Discrete Fourier Transform' (DFT). They can be associated with the true Fourier transform by the following argument:

Suppose, as usual, that $f(x)$ and $\phi(p)$ are a Fourier pair. If $f(x)$ is multiplied by a III-function of period a then the Fourier transform becomes:

$$\int_{-\infty}^{\infty} f(x) III_a(x)e^{2\pi ipx}dx = \frac{1}{a}\phi(p) * III_{1/a}(p)$$

Now suppose that $f(x)$ is negligibly small for all x outside the limits $-a/2 \rightarrow (N-1/2)a$, so that there are N teeth in the Dirac comb, and $f(x)$ extends over a range $\leq Na$. We rewrite the integral and use the properties of δ-functions so that

$$\Phi(p) = \int_{-\infty}^{\infty} \sum_{n=-\infty}^{\infty} f(x)e^{2\pi ipx}\delta(x-na)dx$$

$$= \sum_{n=-\infty}^{\infty} \int_{-\infty}^{\infty} f(x)e^{2\pi ipx}\delta(x-na)dx$$

and because there are only N teeth in the comb the integral has finite limits, and

$$\Phi(p) = \sum_{n=0}^{N-1} f(na)e^{2\pi ipna}$$

and this sum is $(1/a)[\phi(p) * III_{\frac{1}{a}}(p)]$.

† J.W. Cooley & J.W. Tukey, 'An algorithm for the machine calculation of complex Fourier series', *Math. Computation*, **19** 297–301, April 1965.

This in turn, is periodic in p with period $1/a$, and can be written:

$$\Phi(p) = (1/a)\phi(p) * III_{\frac{1}{a}}(p) =$$
$$(1/a)[\phi(p) + \phi(p + 1/a) + \phi(p - 1/a) + \phi(p + 2/a) + \phi(p - 2/a) + \ldots)$$

and the fundamental period of $\Phi(p)$ is the same as $\phi(p)$ if $\phi(p)$ occupies only the range $\pm 1/2a$.

Now consider n intervals of p each of width $1/Na$. At the mth interval the equation becomes:

$$\Phi(m/Na) = \sum_{n=0}^{N-1} f(na)e^{2\pi ina(m/Na)} = (1/a)\phi(m/Na)$$

or, more succinctly:

$$\sum_{n=0}^{N-1} f(n)e^{2\pi inm/N} = (1/a)\phi(m)$$

and this approximates to the analytic Fourier transform. The approximation is that $\Phi(p) = \phi(p)$ in its first period. Theoretically it is not – there is bound to be some overlap since $\phi(p)$ is not zero – but practically it can be ignored†.

The choice of the interval $-a/2 \rightarrow (N - 1/2)a$ for $f(x)$ is so as to have exactly N teeth in the Dirac comb without the embarrassment of having teeth at the very edge – where a top-hat function changes from 1 to 0, for example. In theory *any* interval of the same length would do.

9.3 The matrix form of the DFT

One way of looking at the formula for the Discrete Fourier Transform is to set it out as a matrix operation. The data set $[a(n)]$ can be written as a column matrix or 'vector' (in an n-dimensional space), to be multiplied by a square matrix containing all the exponentials and giving another

---

† It is not possible for a function and its Fourier pair both to be finite in extent – one at least must extend to $\pm\infty$ but the condition that both be small compared with the values in the region of interest is allowable.

column matrix with n components, $[A(m)]$ as its result.

$$
\begin{vmatrix}
A(0) \\
A(1) \\
A(2) \\
A(3) \\
\vdots \\
A(N-1)
\end{vmatrix}
=
\begin{Vmatrix}
1 & 1 & 1 & \ldots & 1 \\
1 & e^{2\pi i/n} & e^{4\pi i/n} & \ldots & e^{2(N-1)\pi i/n} \\
1 & e^{4\pi i/n} & e^{8\pi i/n} & \ldots & e^{4(N-1)\pi i/n} \\
1 & e^{6\pi i/n} & e^{12\pi i/n} & \ldots & e^{6(N-1)\pi i/n} \\
\vdots & \vdots & \vdots & \vdots & \vdots \\
1 & \ldots & \ldots & \ldots & e^{(N-1)^2\pi i/n}
\end{Vmatrix}
\begin{vmatrix}
a(0) \\
a(1) \\
a(2) \\
a(3) \\
\vdots \\
a(N-1)
\end{vmatrix}
\qquad (9.2)
$$

The process of matrix multiplicaton requries n^2 multiplications for its completion. If large amounts of data are to be processed, this can become inordinate, even for a computer. Some people like to process columns of data with 10^6 numbers occasionally, but normally experimenters make do with 1024, although they require the transform in a millisecond or less.

The secret of the Fast Fourier Transform is that it reduces the number of multiplications to be done from N^2 to $2N\log_2(N)$. A data 'vector' 10^6 numbers long then requires 4.2×10^6 multiplications instead of 10^{12}, a gain in speed of approximately $\times 262\,000$. In this year of grace 1994, the computation time is reduced from hours to a few tens of milliseconds.

The way it does this is, in essence, to factorise the matrix of exponentials, but there are easier ways of looking at the process. For example: suppose that the number N of components in the vector is the product of two numbers k and l. Instead of writing the subscript of each number in the vector to denote its position $(0 \ldots N-1)$ it can be given two subscripts s and t, and written $a(s,t)$, $= a(sk + t)$, where s takes values from 0 to $(l-1)$ and t runs from 0 to $(k-1)$. In this way all the numbers in the vector are labelled, but now with two suffixes instead of one. There is absolutely no point in doing this except for computational purposes: it is purely a piece of computer-mathematical manipulation, and would have struck mathematicians of pre-computer days as ludicrous. However, we now write the Digital transform as:

$$
A(u,v) = \sum_{s=0}^{l-1} \sum_{t=0}^{k-1} a(s,t) e^{2\pi i(sk+t)(ul+v)/kl}
$$

where the suffix m in the transformed vector has similarly been dissected into u and v, with $m = ul + v$. u runs from 0 to $(k-1)$ and v from 0 to $(l-1)$.

The exponent is now multiplied out and gives

$$A(u,v) = \sum_{s=0}^{l-1} \sum_{t=0}^{k-1} a(s,t) e^{2\pi i s u} e^{2\pi i s v/l} e^{2\pi i t u/k} e^{2\pi i v t/kl}$$

The first exponential factor is unity and is discarded. The double sum can be rewritten now as:

$$A(u,v) = \sum_{t=0}^{l-1} e^{2\pi i t u/k} e^{2\pi i v t/kl} \sum_{s=0}^{k-1} a(s,t) e^{2\pi i s v/l}$$

which is legitimate since only the last exponent contains a factor s.

This sum over k terms gives a new set of numbers $[g(v,t)]$ and we write:

$$A(u,v) = \sum_{t=0}^{l-1} [g(v,t) e^{2\pi i v t/kl}] e^{2\pi i t u/k}$$

The array $[g(v,t)]$ is multiplied by $[e^{2\pi i v t/kl}]$ to give an array $[g'(v,t)]$ and finally the sum:

$$g''(v,u) = \sum_{t=0}^{l-1} g'(v,t) e^{2\pi i t u/k}$$

and $g''(v,u) = A(u,v)$. (The reversing of the order of v and u is important.)

The transform has been split into two stages: There are k transforms, each of length l, followed by N multiplications by the exponential factors $e^{2\pi i v t/kl}$ (the 'twiddle-factors'), followed by l transforms, each of length k: a total of $kl^2 + lk^2 = N(k+l)$ multiplications, apart from the relatively small number, N of multiplications [by $e^{2\pi i v t/kl}$] in the middle.

The lesson is that, provided N can be factorized, the vector $[a(n)]$ can be turned into a rectangular $k \times l$ matrix and treated column by column as a set of shorter transforms. For example, if there were a factor 2, the *even* numbered a's could be put into one vector of length $N/2$ and the odd-numbered a's into another. Then each is subjected to a Fourier transform of half the length to give two more vectors, and these, after multiplying by the so-called 'twiddle-factors' as above, can be recombined into a vector of length N.

The same process can be repeated provided that $N/2$ can be factored; and if the factors are always 2, it continues until only 2×2 matrices are left, with trivially easy Fourier transforms (and a multiplicity of twiddle-factors!) The interesting thing is that each number in the transformed vector has its address in bit-reversed order. In the example given earlier

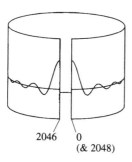
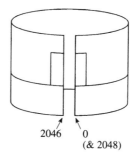

2046   0
(& 2048)

2046   0
(& 2048)

Fig. 9.1. the implementation of the FFT using a sinc-function as an example. The two cylinders unwrapped, represent the input and output data arrays. Do not expect zero to be in the middle as in the analytic case of a Fourier transform. If the input data are symmetrical about the centre, these two halves must be exchanged (*en bloc*, not mirror-imaged) before and after doing the FFT.

the final outcome was $g''(v, u)$, so that the two indices have to be reversed – the number $g''(v, u)$ is in the wrong place in the array. This effect is multiplied until, in the 2^N transform, the transformed data appear in the wrong addresses, the true address being the bit-reversed order of the apparent address.

The Fast Fourier Transform is thus usually done with N a power of 2. Not only is it very efficient in terms of computing time, but is ideally suited to the binary arithmetic of digital computers. The details of the way programs are written are given by Brigham† and a BASIC listing of an FFT routine is given at the end of this chapter. There are many such routines, the results of many hours of research, and sometimes very efficient. This one is not particularly fast but will suffice for practice and is certainly suitable for student laboratory work.

The data file for this program must be 2048 words long (1024 complex numbers, alternately real and imaginary parts), and if only real data are to be transformed, they should go in the even numbered elements of the array, from 0 to 2046. Some caution is needed: zero frequency is at array element 0. If you want to Fourier transform a sinc-function for example, the positive part of the function should go at the beginning of the array and the negative part at the end. The diagram illustrates the point: the output will similarly contain the zero-frequency value in element 0, so that the top-hat appears to be split between the beginning and the end.

Alternatively, you can arrange to have zero-frequency at point 1024

† Brigham, loc.cit.

in the array, in which case the input and output arrays must both be transposed, by having the first and second halves interchanged (but not flipped over) before and after the FFT is done.

Attention to these details saves a lot of confusion! It helps to think of the array as wrapped around a cylinder, with the beginning of the array at zero-frequency and the end at point (-1) instead of $(+1023)$.

The output can be used in a straightforward way to give the power, phase or modular transforms, and the data can be presented graphically with simple routines which need no description here.

9.4 The BASIC FFT routine

The listing below is of a simple BASIC routine for the Fast Fourier Transform of 1024 complex numbers.

This is a routine which can be incorporated into a program which you can write for yourself.

The data to be transformed are put in an array $D(I)$ declared at the beginning of the program as 'DIM D(2047)', and the reals go in the even numbered places, beginning at 0, and the imaginaries in the odd-numbered. The transformed data are found similarly in the same array. The variable G on line 131 should be set to -1 for a direct transform and to 1 for an inverse transform. Numbers to be entered into the $D(I)$ array should be in ASCII format. The program should fill the $D(I)$ array with data; call the FFT as a routine with a 'GOSUB 100' statement (The 'RETURN' is the last statement, on line 159). and this can be followed by instructions for displaying the data.

It is well worth while incorporating a routine for transposing the two halves of the $D(I)$ array before and after doing the transform, as an aid to understanding what is happening.

```
(instruction 'GOSUB 100' gets you to this point)
100 QQ = 1024
101 Z = 2048
102 PRINT 'BEGIN FFT'
103 LET N =Z
104 LET J =1
105 FOR I = 1 TO N STEP 2
106 IF (I-J)<0 GOTO 109
107 IF I= J GOTO 115
108 IF (I-J)>0 GOTO 115
```

```
109 LET T=D(J-1)
110 LET S= D(J)
111 LET D(J-1)=D(I-1)
112 LET D(J)=D(I)
113 LET D(I-1)=T
114 LET D(I)=S
115 LET M=N/2
116 IF (J-M)<0 GOTO 124
117 IF J=M GOTO 124
118 IF (J-M)>0 GOTO 119
119 LET J=J-M
120 LET M=M/2
121 IF (M-2)<0 GOTO 124
122 IF M=2 GOTO 116
123 IF(M-2)>0 GOTO 116
124 LET J=J+M
125 NEXT I
126 LET X=2
127 IF (X-N)<0 GOTO 130
128 IF X=N GOTO 153
129 IF (X-N)>0 GOTO 153
130 LET F=2*X
131 LET H= -6.28319/(G*X)
132 LET R= SIN(H/2)
133 LET W=-2*R*R
134 LET V= SIN(H)
135 LET P=1
136 LET Q=0
137 FOR M=1 TO X STEP 2
138 FOR I=M TO N STEP F
139 LET J=I+X
140 LET T= P*D(J-1)-Q*D(J)
141 LET S= P*D(J)+Q*D(J-1)
142 LET D(J-1)=D(I-1)-T
143 LET D(J)=D(I)-S
144 LET D(I-1)=D(I-1)+T
145 LET D(I)=D(I)+S
146 NEXT I
147 LET T=P
148 LET P=P*W-Q*V+P
```

```
149 LET Q=Q*W+T*V+Q
150 NEXT M
151 LET X=F
152 GOTO 128
153 CLS
154 REM
155 FOR I = 0 TO 2047
156 LET D(I)=D(I)/32
157 NEXT I
158 PRINT 'FFT DONE'
159 RETURN
```

And here is a short program to generate a file with DAT extension which will contain a top-hat function of any width you choose. The data are generated in ASCII and can be used directly with the FFT program above.

```
1 REM Program to generate a 'Top-hat' function.
2 INPUT 'input desired file name', A$
3 INPUT 'Top-hat Half-width ?', N
4 PI = 3.141592654
5 DIM B(2047)
6 FOR I = 1024-N TO 1024+N STEP 2
7 B(I) = 1/(2 * N)
8 NEXT I
9 C$='.DAT'
10 C$=A$+C$
11 PRINT
12 OPEN C$ FOR OUTPUT AS #1
13 FOR I = 0 TO 2047
14 PRINT #1,B(I)
15 NEXT I
16 CLOSE #1
```

the simple file-generating arithmetic in lines 6–8 can obviously be replaced by something else, and this sort of 'experiment' is of great help in understanding the Fast Fourier Transform process.

The file thus generated can be read into the FFT program with:

```
20 REM SUBROUTINE FILELOAD
21 REM TO OPEN A FILE AND LOAD CONTENTS INTO D(I)
22 GOSUB 24
```

```
23 (insert the next stage of the program, e.g.'gosub 100', here)
24 CLS:LOCATE 10,26,0
25 PRINT' NAME OF DATA FILE ?'
26 LOCATE 14,26,0
27 INPUT A$
28 ON ERROR GOTO 35
29 OPEN 'I',#1,A$
30 FOR I = 0 TO 2047
31 ON ERROR GOTO 35
32 INPUT#1,D(I)
33 NEXT I
34 CLOSE
35 RETURN
```

Appendix 1

A1.1 The Heaviside step-function

This has the properties that:

$$H(x) = 0, X < 0 \text{ and } H(x) = 1, x > 0$$

and it is convenient to assume that $H(0) = \frac{1}{2}$.

Its Fourier transform is obtained easily. It can be regarded as the integral of the δ-function and the integral theorem (q.v.) can be used to derive it.

$$\partial H(x)/\partial x = \delta(x)$$

$$\partial H(x)/\partial x \rightleftharpoons 1$$

$$H(x) \rightleftharpoons 1/2\pi i p$$

It can be manipulated in the usual way:

$$-H(x - a/2) = -H(x) * (\delta(x - a/2))$$

and

$$H(x - a/2) - H(x + a/2) = H(x) * [\delta(x + a/2) - \delta(x - a/2)]$$

and the Fourier transform of the right hand side is:

$$\frac{-1}{2\pi i p}[e^{-i\pi pa} - e^{i\pi pa}]$$

$$\tfrac{1}{\pi p}\sin \pi pa = a\,\mathrm{sinc}\pi pa$$

and the left hand side is clearly a top-hat function of width a.

It is used chiefly much as a top-hat function is used, to isolate parts of another function. For example a sinusoidal wave switched on at time $t = 0$ can be written as $f(t) = \cos 2\pi v t . H(t)$.

A1.2 Parseval's Theorem and Rayleigh's Theorem

Parseval's Theorem states that:

$$\int_{-\infty}^{\infty} f(x)g^*(x)dx = \int_{-\infty}^{\infty} F(p)G^*(p)dp$$

This proof relies on the fact that if

$$g(x) = \int_{-\infty}^{\infty} G(p)e^{2\pi ipx}dp$$

then

$$g^*(x) = \int_{-\infty}^{\infty} G^*(p)e^{-2\pi ipx}dp$$

(simply by taking complex conjugates of everything).

Then it follows that:

$$G^*(p) = \int_{-\infty}^{\infty} g^*(x)e^{2\pi ipx}dx$$

The left hand side can be written as:

$$f(x)g^*(x) = \int_{-\infty}^{\infty} F(q)e^{2\pi iqx}dq \int_{-\infty}^{\infty} G^*(p)e^{-2\pi ipx}dp$$

We integrate both sides with respect to x. If we choose the order of integration carefully, we find:

$$\int_{-\infty}^{\infty} f(x)g^*(x)dx = \int_{-\infty}^{\infty} \left\{ \int_{-\infty}^{\infty} F(q) \left[\int_{-\infty}^{\infty} G^*(p)e^{-2\pi ipx}dp \right] e^{2\pi iqx}dq \right\} dx$$

and changing the order of integration:

$$= \int_{-\infty}^{\infty} \left\{ F(q) \int_{-\infty}^{\infty} g^*(x)e^{2\pi iqx}dx \right\} dq$$

$$= \int_{-\infty}^{\infty} F(q)G^*(q)dq$$

The theorem is often seen in a simplified form, with $g(x) = f(x)$ and $G(p) = F(p)$. Then it is called **Rayleigh's Theorem.**

Another version of Parseval's Theorem involves the coefficients of a Fourier *series*. In words, it states that the average value of the square of $F(t)$ over one period is the sum of the squares of all the coefficients of the series.

The proof is as follows:

With the usual assumption, that

$$F(t) = \sum_{-\infty}^{\infty} a_n \cos(2\pi n v_0 t) + b_n \sin(2\pi n v_0 t)$$

and coefficients as before, given by:

$$a_n = \int_0^{1/v_0} F(t) \cos(2\pi n v_0 t) dt; \qquad b_n = \int_0^{1/v_0} F(t) \sin(2\pi n v_0 t) dt$$

the left hand side can be written:

$$\int_0^{1/v_0} F(t)F(t)dt \; = \; \sum_{-\infty}^{\infty} a_n \int_0^{1/v_0} F(t) \cos(2\pi n v_0 t) dt + b_n \int_0^{1/v_0} \sin(2\pi n v_0) dt$$

$$= \; \sum_{-\infty}^{\infty} \left(\frac{a_n^2}{v_0} + \frac{b_n^2}{v_0} \right)$$

and then

$$v_0 \int_0^{1/v_0} \left[F(t)^2 \right] dt = \sum_{-\infty}^{\infty} (a_n^2 + b_n^2)$$

or, using the coefficients for the half-range series ($n = 1 \rightarrow \infty$)

$$v_0 \int_0^{1/v_0} \left[F(t)^2 \right] dt = \frac{A_0^2}{4} + \frac{1}{2} \cdot \sum_1^{\infty} (A_n^2 + B_n^2)$$

A1.3 Useful formulae from Bessel function theory

A1.3.1 The Jacobi expansion

$$e^{ix \cos y} = J_0(x) + 2 \sum_{n=1}^{\infty} i^n J_n(x) \cos ny$$

$$e^{ix \sin y} = \sum_{z=-\infty}^{\infty} J_z(x) e^{izy}$$

A1.3.2 The integral expansion

$$J_0(2\pi\rho r) = \frac{1}{2\pi} \int_0^{2\pi} e^{2\pi i \rho r \cos\theta} d\theta$$

which is a particular case of the general formula:

$$J_n(x) = \frac{i^{-n}}{2\pi} \int_0^{2\pi} e^{in\theta} e^{ix\cos\theta} d\theta$$

$$\frac{d}{dx}\left(x^{n+1}J_{n+1}(x)\right) = x^{n+1}J_n(x)$$

A1.3.3 The Hankel transform

This is similar to a Fourier transform, but in polar coordinates, r, θ. The Bessel functions form a set with orthogonality properties similar to those of the trigonometrical functions and there are similar inversion formulae. These are:

$$F(x) = \int_0^\infty pf(p)J_n(p)dp$$

$$f(p) = \int_0^\infty xF(x)J_n(x)dx$$

where J_n is a Bessel function of any order.

Bessel functions are analogous in many ways to the trigonometric functions sin and cos. In the same way as sin and cos are the solutions of $\frac{d^2y}{dx^2} + k^2x = 0$, they are the solutions of *Bessel's equation*, which is:

$$x^2\frac{d^2y}{dx^2} + x\frac{dy}{dx} + (x^2 - n^2)y = 0$$

In its full glory, n need not be integer and neither x nor n need be real.

The functions are tabulated in various books† for real x and for integer and half-integer n, and can be calculated numerically, as are sines and cosines, by computer.

In its simpler form, as shown, it occurs with θ as variable when Laplace's equation is solved in cylindrical polar coordinates and variables are separated to give functions $R(r)\Theta(\theta)\Phi(\phi)$, and this is why it proves useful in Fourier transforms with circular symmetry.

† For example, in Jahnke & Emde (see bibliography).

A1.4 Conversion of Fourier series coefficients to complex exponential form

We use De Moivre's theorem to do the conversion. Writing $2\pi v_0 t = \theta$ for the moment, we can also write:

$$F(t) = \sum_{m=-\infty}^{\infty} (a_m \cos m\theta + b_m \sin m\theta)$$

and then by De Moivre's theorem:

$$
\begin{aligned}
F(t) &= \sum_{m=-\infty}^{\infty} \frac{a_m}{2}(e^{im\theta} + e^{-im\theta}) + \frac{b_m}{2i}(e^{im\theta} - e^{-im\theta}) \\
&= \sum_{m=-\infty}^{\infty} \frac{a_m - ib_m}{2} e^{im\theta} + \sum_{m=-\infty}^{\infty} \frac{a_m + ib_m}{2} e^{-im\theta}
\end{aligned}
$$

Now put $n = -m$. Then:

$$F(t) = \sum_{m=-\infty}^{\infty} \frac{a_m - ib_m}{2} e^{im\theta} + \sum_{n=-\infty}^{\infty} \frac{a_{-n} + ib_{-n}}{2} e^{in\theta}$$

and since $a_m = a_{-m}$ and $b_m = -b_{-m}$ and the sums run from $-\infty$ to ∞ (meaning that n is a 'dummy' suffix and can be replaced by any other letter; m for example):

$$
\begin{aligned}
F(t) &= \sum_{m=-\infty}^{\infty} e^{im\theta} \left\{ \frac{a_m}{2} - \frac{ib_m}{2} + \frac{a_{-m}}{2} + \frac{ib_{-m}}{2} \right\} \\
&= \sum_{m=-\infty}^{\infty} e^{im\theta} \left\{ \frac{(A_m - iB_m)}{2} \right\} \\
&= \frac{A_0}{2} + \sum_{m=1}^{\infty} e^{im\theta}(A_m - iB_m)
\end{aligned}
$$

or, replacing θ:

$$
\begin{aligned}
F(t) &= \frac{A_0}{2} + \sum_{m=1}^{\infty} e^{2\pi i m v_0 t}(A_m - iB_m) \\
&= \frac{A_0}{2} + \sum_{m=1}^{\infty} e^{2\pi i m v_0 t} C_m
\end{aligned}
$$

as in chapter 1.

Notes and Bibliography

The most popular books on the practical applications of Fourier theory are undoubtedly those of Champeney and Bracewell and they cover the present ground more thoroughly and in much more detail than here. E. Oran Brigham, on the Fast Fourier Transform, is the classic work on the subjects dealt with in chapter 9.

Of the more theoretical works, the 'bible' is Titchmarsh, but a more readable (and entertaining) work is Körner's. Whittaker's (not to be confused with the more prolific E. T. Whittaker) is a specialised work on interpolation, but that is a subject which is getting more and more important, especially in computer graphics.

Many writers on Quantum Mechanics, Atomic Physics and Electronic Engineering like to include an early chapter on Fourier theory. One or two (who shall be nameless) get it wrong! They confuse ω with v or leave out a 2π when there should be one, or something like that. The specialist books, like those below, are much to be preferred.

Abramowitz, M. & Stegun, I.A. *Handbook of Mathematical Functions.* Dover, New York. 1965.
  A more up-to-date version of Jahnke & Emde, below.
Bracewell, R.N. *The Fourier Transform and its Applications.* McGraw-Hill, New York. 1965.
  This is one of the two most popular books on the subject. Similar in scope to this book, but more thorough and comprehensive.
Brigham, E.O. *The Fast Fourier Transform.* Prentice Hall, New York. 1974.
  The standard work on digital Fourier transforms and their implementation by various kinds of FFT programs.
Champeney, D.C. *Fourier Transforms and Their Physical Applications.* Academic Press, London & New York. 1973.
  Like Bracewell, one of the two most popular books on practical Fourier tranforming. Covers similar ground, but with some differences.
Champeney, D.C. *A Handbook of Fourier Theorems.* Cambridge University Press. 1987.

Jahnke, E. & Emde, F. *Tables of Functions with Formulae and Curves*. Dover, New York. 1943.
The great classic work on the functions of mathematical physics, with diagrams, charts and tables, of Bessel functions, Legendre polynomials etc.

Körner, T.W. *Fourier Analysis*. Cambridge University Press. 1988.
One of the more thorough and entertaining works on analytic Fourier theory, but plenty of physical applications: expensive, but firmly recommended for serious students.

Titchmarsh, E.C. *An Introduction to the Theory of Fourier Integrals*. Clarendon Press, Oxford. 1962.
The theorists' standard work on Fourier theory. Unnecessarily difficult for ordinary mortals, but needs consulting occasionally.

Watson, G.N. *A Treatise of the Theory of Bessel Functions*. Cambridge University Press. 1962.
Another great theoretical classic: chiefly for consultation by people who have equations they can't solve, and which seem likely to involve Bessel functions.

Whittaker, J.M. *Interpolary Function Theory*. Cambridge University Press. 1935.
A slim volume dealing with the problems of interpolating points between samples of band-limited curves.

Index